Walter Grünzweig (Ed.)

The SciArtist

Transnational and Transatlantic American Studies

edited by

Kornelia Freitag (Bochum)
Walter Grünzweig (Dortmund)
Randi Gunzenhäuser (Dortmund)
Martina Pfeiler (Dortmund)
Wilfried Raussert (Bielefeld)
Michael Wala (Bochum)

Volume 11

The SciArtist

Carl Djerassi's Science-in-Literature
in Transatlantic
and Interdisciplinary Contexts

edited by

Walter Grünzweig

LIT

Cover Art / Umschlagbild:
Gabriele Seethaler, 2012, *Gruppe zu elf mutant 3 übermalt, überlagert von Carl Djerassi*; nach Paul Klees *Gruppe zu elf*

Gedruckt auf alterungsbeständigem Werkdruckpapier entsprechend
ANSI Z3948   DIN ISO 9706

Bibliographic information published by the Deutsche Nationalbibliothek
The Deutsche Nationalbibliothek lists this publication in the Deutsche Nationalbibliografie; detailed bibliographic data are available in the Internet at http://dnb.d-nb.de.

ISBN 978-3-643-90231-3

A catalogue record for this book is available from the British Library

©**L**IT VERLAG GmbH & Co. KG Wien,
Zweigniederlassung Zürich 2012
Klosbachstr. 107
CH-8032 Zürich
Tel. +41 (0) 44-251 75 05
Fax +41 (0) 44-251 75 06
e-Mail: zuerich@lit-verlag.ch
http://www.lit-verlag.ch

LIT VERLAG Dr. W. Hopf
Berlin 2012
Fresnostr. 2
D-48159 Münster
Tel. +49 (0) 2 51-620 320
Fax +49 (0) 2 51-23 19 72
e-Mail: lit@lit-verlag.de
http://www.lit-verlag.de

Distribution:
In Germany: LIT Verlag Fresnostr. 2, D-48159 Münster
Tel. +49 (0) 2 51-620 32 22, Fax +49 (0) 2 51-922 60 99, e-mail: vertrieb@lit-verlag.de

In Austria: Medienlogistik Pichler-ÖBZ, e-mail: mlo@medien-logistik.at

In Switzerland: B + M Buch- und Medienvertrieb, e-mail: order@buch-medien.ch

In the UK: Global Book Marketing, e-mail: mo@centralbooks.com

In memoriam Petra Meurer
(1968 – 2010)

Contents

Acknowledgment	1
Geleitwort: Naturwissenschaft – Kulturwissenschaft – Wissenschaftskultur. Carl Djerassi und der interdisziplinäre Entwurf der Welt *Matthias Kleiner*	3
The Renaissance Man	7
The Lives of Carl Djerassi *David Lodge*	9
Mehr als ein „Renaissance Man": Die Präsenz von Carl Djerassi *Hans Ulrich Gumbrecht*	15
The Two/Several Cultures	23
Cross Talk: A Dialog between Carl Djerassi and George Klein *Carl Djerassi and George Klein*	25
Weak Bonds – Strong Effects: How Carl Djerassi Left Chemistry and Became a Writer *Henning Hopf*	43
Carl Djerassi: Scientist – Artist – SciArtist *Christian R. Noe*	53
Science-in-Theatre	61
Balancing Act: Drama about Science based on History *Robert Marc Friedman*	63
"Ambition without Love is Cold": Priority and Kudos in *Oxygen* by Carl Djerassi and Roald Hoffmann *Eva-Sabine Zehelein*	75

Carl Djerassi: U.S.-amerikanischer Dramatiker Wiener Herkunft auf dem
Weg nach Europa 85
Isabella Gregor

Carl Djerassi's Science-in-Theatre Plays: The Theatrical Realization 97
Andy Jordan

Science-in-Fiction 133

"If something like a next life exists, yes, I'd rather be a woman": Carl
Djerassi's Science-in-Fiction and the Regendering of the Natural Sciences 135
Ingrid Gehrke

The Academic Novel: A Personal Typology 145
Pierre Laszlo

Science-in-Poetry 155

Private Words Addressed in Public: Carl Djerassi's Poetry at the
Science-to-Humanities Interface 157
Federico Maria Rubino

Science-in-Poetry? Chemistry and the Metaphysical Tradition 169
Walter Grünzweig

Arts and Translation 183

Multiplizität als künstlerische Strategie:
Zum wissenschaftlich-künstlerischen Fotodialog von Gabriele Seethaler
und Carl Djerassi 185
Carl Aigner

Carl Djerassi und Paul Klee: Ein Dramolett 193
Klaus Albrecht Schröder

From the Pill to Paul Klee: Translating Carl Djerassi 199
Ursula-Maria Mössner

Parnassus 207

Four Jews on Parnassus: Carl Djerassi's Refiguration of the
German-Jewish Parnassus 209
Erin McGlothlin

The Idle Chatter of Ghosts: On Carl Djerassi's *Four Jews on Parnassus* Martin Jay	221

Carl Djerassi: A Literary Bibliography 231

Contributors	249
Index	255

Acknowledgment

The editor wishes to thank the many people who made this first volume of Djerassi criticism possible – foremost, of course, the contributors, the German Research Foundation, and TU Dortmund University. We thank all those who provided artistic, aesthetic and media expertise, most notably Astrid Kämmerling, Gabriele Seethaler and Florian Siedlarek. Special intellectual and research advice was provided by Eva-Sabine Zehelein. Support for all aspects of the book came from Carl Djerassi himself. Our warmest thanks go to our benefactors Dietrich Groh and Barbara Blümel who have not only taken an active interest in Carl Djerassi's work but who have also generously supported the publication of this book financially.

Geleitwort:
Naturwissenschaft – Kulturwissenschaft – Wissenschaftskultur.
Carl Djerassi und der interdisziplinäre Entwurf der Welt

Matthias Kleiner

Im Jahr 2009 erhielt Carl Djerassi an der Technischen Universität Dortmund nach zwanzig Ehrentiteln für seine wissenschaftlichen Leistungen in der Chemie die bislang einzige Ehrendoktorwürde für sein literarisches Werk, die erste akademische Auszeichnung in seiner zweiten Karriere. Für mich als Präsident der Deutschen Forschungsgemeinschaft, die das mit dem Ehrendoktorat verbundene Symposium zum literarischen Werk Carl Djerassis gefördert hat, ist diese Anerkennung besonders bedeutsam, da sie den interdisziplinären Bereich zwischen Kultur und Technik anspricht, der für die Arbeit DFG sehr wichtig ist. Das oft mangelnde Interesse am Ingenieurstudium hat auch mit der Entfremdung der beiden zusammengehörenden Bereiche von Kultur und Technik zu tun. Jede ernstzunehmende technische Universität wird aus wohlverstandenem Eigeninteresse, aber auch im Interesse des universitären Ganzen, seine kultur- und geisteswissenschaftlichen Fächer ganz besonders fördern und sie nicht im Bereich der Zweitklassigkeit marginalisieren.

Djerassis Beispiel ist für diese Zusammenarbeit besonders instruktiv. Während Interdisziplinarität in den meisten Fällen hergestellt wird, indem sich Kolleginnen und Kollegen aus verschiedenen Fächerkulturen und Fachbereichen zusammen finden, um an einem fachübergreifenden Projekt zu arbeiten, stellt Carl Djerassis literarisches Werk sozusagen interfachliche Primärliteratur dar – dieser Autor manifestiert in seiner Person selbst Interdisziplinarität.

Dreiundzwanzig Jahre ist es her, dass Carl Djerassis erster Roman *Cantors Dilemma* in die literarische *und* in die wissenschaftliche Welt fast wie ein Blitz einschlug. Das Neue hier war nicht, dass ein weltberühmter, international hochdekorierter Chemiker bewies, dass er einen Roman schreiben konnte. Das haben andere Chemiker und Naturwissenschaftler auch getan. Das Wesentliche war auch nicht, wie viele Kritiker, auch aus den Naturwissenschaften, behaupteten, dass hier jemand mutig genug war, schmutzige Laboratoriumskittel in aller Öffentlichkeit zu waschen.

Das Neue und Faszinierende war, dass Djerassi zeigte, dass es sich bei den Naturwissenschaften und der Welt der Naturwissenschaften selbst um eine Kultur handelte. Die oftmals gebrauchte, eher metaphorische Wendung von der Wissenschaftskultur erhielt hier eine konkrete Basis in Form einer Romanwelt. *Stammesgeheimnisse*, so nannte Carl Djerassi die ersten beiden Teile seiner Roman-Tetralogie, und gibt den Romanen so einen anthropologischen Auftrag, nämlich die Qualität dieser Kultur darzustellen, die bislang ja wirklich nicht explizit beschrieben worden ist.

Während der Roman eine Welt erzählt, stellt das Theaterstück sie auf die Bühne. Vermutlich erlangte Carl Djerassi im Laufe seiner Arbeit als Romancier die Einsicht, dass man diese Welt der Naturwissenschaften – und nicht nur der Naturwissenschaften – nicht bloß als Diskurs fassen, als Narrativ erzählen kann, sondern dass man sie noch besser versteht, wenn man die Akteure in ihren *Rollen* begreift.

Wissenschaftler als Schauspieler in hoch verwickelten dramatischen Handlungen, das ist es, was Carl Djerassi uns in vielen seiner Theaterstücke bietet. Die Wissenschaft ist selbst dramatisch. Oft definiert sie ein Problem, entwickelt es, läuft auf eine Krise hinaus und steuert schließlich, mit einigen retardierenden Momenten, auf eine Lösung zu.

Zu Djerassis Stück *Taboos*, das von Studierenden der renommierten Fakultät Kulturwissenschaften der TU Dortmund unter der Leitung der Amerikanistin Randi Gunzenhäuser und der tragisch früh verstorbenen Dortmunder Germanistin Petra Meurer anlässlich der Verleihung Ehrendoktorats in Dortmund und anlässlich der Dreihundertjahrfeier der Charité in Berlin aufgeführt wurde, schrieb Neil Genzlinger am 24. September 2008 in der *New York Times*:

Carl Djerassi hat eine ganze Menge Schwangerschaften verhütet, aber in seinem Stück *Taboos* ereignet sich eine Mini-Bevölkerungsexplosion. Es bietet Mr. Djerassi eine Möglichkeit, sich dramatisch dem neuen emotionalen Terrain zu nähern, das die Wissenschaft durch die künstliche Befruchtung und andere Reproduktionstechniken eröffnet hat. Romantische Liebe, Elternliebe, biologische Beziehungen, Familie, Heirat: diese und andere Themen tauchen auf, wenn die Handlung von *Taboos* sich entwickelt. Das Stück ist zwar keine Komödie, hat aber einige wunderbare humorvollen Momente.

Es sind alte, dem Theater wohl vertraute Begriffe, die Carl Djerassi hier aufgreift: Familie, Beziehungen, Liebe, Elternschaft. Aber er stellt sie in den neuen Kontext der Möglichkeiten, die die modernen Technologien uns eröffnet haben, die wir kritisieren oder willkommen heißen wollen, die aber einfach da sind und mit denen wir umgehen lernen müssen.

Folgt man der Kritik und auch den Reaktionen von Lesern und Leserinnen, so scheint Djerassi hier die Grundlagen dessen, was bislang unser menschliches Zusammenleben ausgemacht hat, zunächst einmal auf den Prüfstand zu stellen. Und es gehört sicher vieles auf den Prüfstand, inklusiver alter, lieb gewordener ideologischer Polaritäten. Die Gegensätzlichkeit zwischen dem lesbischen Paar aus Kalifornien und den südstaatlich-fundamentalistischen Eheleuten wird vor der reproduktionstechnischen Thematik stark relativiert.

Aber ich behaupte noch mehr, dass die Herausforderung, vor die uns diese Technologien stellen, uns nicht nur dazu zwingt, die Frage nach Elternschaft, Liebe und Beziehungen neu zu stellen. Sie infundiert diese vielleicht allzu selbstverständlich, allzu blass gewordenen Konzepte mit neuer Energie, mit neuem Leben. Naturwissenschaft und Technik erfordert so die Überprüfung tradierter menschlicher Werte.

Carl Djerassi ist ein U.S.-amerikanischer Chemiker mit europäischen Wurzeln, der vor allem durch die erste Synthese eines oralen Kontrazeptivums und des Kortisons bekannt wurde. Der „Vater der Pille", wie er häufig zu seinem eigenen Unwillen genannt wird, begann in den achtziger Jahren des vorigen Jahrhunderts mit einer zweiten, sehr erfolgreichen Karriere als Schriftsteller. Neben einigen Kurzgeschichten und Gedichten wurde er vor allem durch fünf Romane, von denen vier im Bereich der Naturwissenschaften spielen („Science-in-Fiction"), sowie durch eine Reihe von Theaterstücken („Science-in-Theatre") bekannt. In seinem Dokudrama betitelt *Vier Juden auf dem Parnass* (2008), wendet er sich verstärkt auch Fragen der jüdischen Identität und dessen Rolle in Wissenschaft, Kultur und Gesellschaft zu.

Obwohl Djerassis Texte außerordentlich erfolgreich sind und in den USA, Deutschland und vielen Ländern intensiv rezipiert wurden, gab es nur sehr wenige wissenschaftliche Beiträge zum literarischen Werk des Autors, von denen die meisten lediglich den Charakter von Würdigungen hatten. Die erste Dissertation zum literarischen Werk Djerassis von Ingrid Gehrke wurde 2008 unter dem charakteristischen Titel *Der intellektuelle Polygamist. Carl Djerassis Grenzgänge in Autobiographie, Roman und Drama* publiziert; in eine Habilitationsschrift fand Djerassi erstmals im Jahr 2009 in Eva-Sabine Zeheleins *Science: Dramatic: Science Plays in America and Great Britain* Eingang.

Es ist daher sehr zu begrüßen, dass der vorliegende Band diesen „Grenzgänger" zwischen Wissenschaften, Sprachen, Kulturen und Genres von verschiedenen Perspektiven aus untersucht. Ausgangspunkt ist dabei die Einsicht, dass Djerassis Werk nur *international* bzw. in besonderer Weise *bilateral* (USA/Deutschland, deutschsprachige Länder) verstanden werden kann. Dies ergibt sich einerseits aufgrund seiner transatlantischen Biografie (geb. 1923 in Wien, 1938 Flucht nach Bulgarien, danach in die USA; langjährige Forschungstätigkeit in Mexiko und den USA, seit den neunziger Jahren verstärkte Rückwendung nach Europa), andererseits aufgrund seiner Rezeption, die sich international, vor allem jedoch in den deutschsprachigen Ländern und den USA vollzieht. Die deutschsprachige Rezeption hat dabei einen besonderen Status, da eine Reihe von Werken zuerst in der (von Djerassi genau beobachteten) deutschen Übersetzung und erst danach im englischen Original publiziert wurden. Übersetzungsprobleme werden hier als text- und interpretationsrelevante Dimensionen gesehen, die Verständnis für die grenzgängerischen Lebenswelten der Texte Djerassis eröffnen können.

Von besonderem Interesse ist in diesen Beiträgen zu Djerassis Werk das Interface zwischen *Literatur/Kultur* einerseits und *Naturwissenschaft/Technik* andererseits. Durch seine fiktionalen Analysen der naturwissenschaftlichen Diskurse hat Djerassi das Gegensatzpaar der „zwei Kulturen" von C. P. Snow überwunden und eine neue Gesprächsbasis zwischen diesen beiden Varianten der Aneignung und Interpretation von Welt geschaffen. Der Braunschweiger Organiker Henning Hopf nähert die Chemie in ihrem „nexus of rationality and passion" den Geisteswissenschaften an. Einen weiteren *turn* geben Beiträge, die die Genderdimension der Naturwissenschaft bzw. die jüdischen Traditionen im Werk Djerassis untersuchen.

Carl Djerassis Werke haben immer wieder Diskussionen und intellektuelle Debatten provoziert. Der vorliegende Band nimmt diese fruchtbaren Kontroversen auf und setzt sie – auf höherer Ebene – fort. Er beweist damit einmal mehr die Bedeutung von Djerassis vielstimmigem Werk für das Verständnis unserer komplexen – und damit auch interessanten – Welt.

The Renaissance Man

The Lives of Carl Djerassi

David Lodge

Carl Djerassi seems to have had more than his fair share of lives, some lived sequentially, others simultaneously: world famous research chemist, innovative teacher, birth control guru, successful businessman, art collector, patron of artists, poet, novelist and playwright. That's nine, and I have probably left out some. We have a proverbial saying in English, "a cat has nine lives," and Carl Djerassi has shown, as well as versatility, a remarkable capacity for survival. Crucially, as a young Jewish teenager brought up in Vienna, he qualified for emigration to America just in time to escape the horrors of the Holocaust. Subsequently he overcame several life-threatening or disabling accidents and illnesses, and also tragedies in his personal life – the suicide of his daughter, about which he has written movingly and candidly, and more recently the death of his beloved third wife, Diane Middlebrook, after a long and courageous struggle against cancer. None of these setbacks halted his tireless productivity for long. A few years ago he hopped around the world on crutches, as a broken hip joint mended, monitoring productions of his plays, and doing the research for his dramatic work *Four Jews on Parnassus*. In 2013, he will be ninety.

"Workaholic" doesn't begin to describe this man's creative energy, ambition, and industry. He has published over twelve hundred articles and seven monographs on organic chemistry, and it would take several minutes to recite the names of all the honors, awards and honorary degrees he has received for his achievements in pure and applied science. He has published numerous books for the general reader, including two fascinating volumes of autobiography, six works of prose fiction, and the texts of seven plays which have been produced all around the world on stage and radio. Famously, while still a young man, he led the team at Syntex, a research establishment in Mexico City (until then right off the map of cutting-edge science) which successfully synthesized the contraceptive pill; and by shrewdly investing in the company's stock, and later acting as an executive of its subsidiaries, he made himself a rich man. But he didn't spend his money on yachts – I doubt whether he has ever taken a vacation in the usual sense of the word – he spent it on art.

By judicious purchasing he acquired a valuable collection of early twentieth century art and then sold it to pay for the conversion of his ranch outside San Francisco into a retreat for artists of all kinds. He built up a unique collection of the work of Paul Klee and gifted it in equal parts to the San Francisco Museum of Modern Art and the Albertina Museum in Vienna. His elegant apartments in San Francisco and London are enhanced by the work, often specially commissioned, of many contemporary artists. When he married the poet, critic, and biographer Diane Middlebrook in 1985 the wedding feast, held at his ranch, featured music for three flutes and soprano specially composed by John Adams and a wedding dance created by the distinguished choreographer Rhonda Martyn and performed by her troupe of dancers. His latest publication, *Four Jews on Parnassus,* is a mixed-media work of art in its own right, a stunning piece of bookmaking comprising text, striking illustrations executed by Gabriele Seethaler (some in collaboration with Carl) and, slipped inside the back cover, a CD of music by Schönberg and other composers mentioned in the dialogue between Walter Benjamin, Theodor Adorno, Gershom Scholem and Arnold Schönberg. The extraordinary breadth of high-cultural interests and expertise Carl Djerassi displays in his life and work, embracing both natural science and the humanities, makes him seem like a Renaissance man in modern dress, (though his style of dressing is often dandyishly distinctive too, for reasons he explains amusingly in his first volume of autobiography).

I first met Carl in October 1992 at the Frankfurt Book Fair, where we were both guests of the Swiss publisher Gerd Haffmans, who was publishing that autumn German editions of my novel *Nice Work* [*Saubere Arbeit*] and Carl's autobiography, *The Pill, Pygmy Chimps and Degas' Horse* [*Die Mutter der Pille*].We were staying at the same hotel, and so saw a good deal of each other over the few days we were there. The Fair is a somewhat intimidating milieu for authors. The huge airless exhibition halls, their miles of aisles filled with publishers' stalls exhibiting thousands and thousands of new books in every language, induce a kind of nausea, or even despair – how can one's work possibly attract attention in that sea of print? And since it is a trade fair, not a readers' fair like Leipzig or Gothenburg, there is not a lot to do except give the occasional interview or have one's photograph taken. So we were glad of each other's company.

It so happened that I had read Carl's first novel, *Cantor's Dilemma,* first published in 1989. On a recent visit to Cambridge, Mass., a Harvard professor I met there had recommended it to me as an excellent academic novel that, unusually, was about scientists. American and British campus novels, from classics like Mary McCarthy's *The Groves of Academe* and Kingsley Amis's *Lucky Jim* onwards, had until then invariably been set in Humanities faculties, for the very good reason that most novelists' experience of university life, as students and/or teachers, is in the humanities; but this meant that fiction was ignoring a very large part of academic life. I bought the American Penguin edition of *Cantor's Dilemma* in a Cambridge bookstore and found it both entertaining and instructive. On my return to England I passed it to my daughter Julia, who was then a post-doc research assistant in microbiology at the University of Birmingham. She enjoyed it too, and said that the presentation of the behavioral aspects of academic science in the novel, especially the power relations between senior and junior staff, and the delicate business of giving individuals credit in the publication of collaborative work, was absolutely spot on. After I told Carl of her opinion he sent me an inscribed copy of his autobiography to give to her.

In the years that followed that meeting in Frankfurt my wife Mary and I became good friends with Carl and Diane. Although we live in Birmingham, I have a *pied à terre* in London which made it easy to accept invitations to dinner parties at the Djerassis's flat in Maida Vale, with its spacious balcony at the back overlooking a small private park shared by the surrounding properties – typical of this part of London, and one of the most civilized forms of urban residence ever devised. The Djerassis – Diane in particular – loved bringing people from different areas of creative and intellectual life together to interact. Every summer we attended the big parties they threw in collaboration with another American anglophile couple, Elaine and English Showalter, to celebrate American Independence Day, when that balcony was crammed with interesting guests from the arts, sciences, literature, journalism, publishing – and the theatre.

My friendship with Carl commenced just around the time when he had decided to try his hand at writing plays as well as novels, and it so happened that I myself had recently made a similar career move. My first play *The Writing Game* [translated into German as *Literatenspiele*] was premiered at the Birmingham Repertory Theatre in 1990, and produced at the American Repertory Theatre in Cambridge, Mass., in 1991, and I was currently engaged in a prolonged, frustrating and ultimately unsuccessful effort to see the play mounted in London and/or New York. So I had a good idea of the difficulties a writer encounters in the essentially collaborative medium of theater, where (compared to the solitary business of novel-writing) so many variables are outside his or her control; and frankly I doubted whether Carl, starting so late, would succeed in overcoming them. I reckoned without his indefatigable determination, ability to learn by experience, and willingness to take infinite pains to succeed.

Surveying Carl Djerassi's remarkable achievements, what most interests me is the shift in the focus of his career from cutting-edge scientific research to writing prose fiction and drama, relatively late in life. I can't think of another example of such a shift that is truly comparable. There have been other writers with scientific qualifications who turned to writing novels – C.P. Snow, for example – but I would venture to say that none of them has been a world-class scientist. The two kinds of activity are so different as to be normally almost incompatible. Science seeks to formulate general laws; literature embodies general truths in the particular. A work of literature consists of certain words in a certain order and cannot be dissociated from them, while, as Carl himself says, "[S]cientific writing aspires to the pure transmission of information: there content is king, style counts for little." (Djerassi 2001, 147) Admittedly some distinguished scientists have written poetry of quality in their spare time, and that was in fact the first kind of creative writing Carl himself tried, but there is a certain affinity between the essentially metaphorical structure of lyric poetry and the ability to perceive connections between apparently unrelated phenomena that underlines so many scientific theories and discoveries. Narrative, the stuff of prose fiction and drama, can incorporate metaphor as trope, but depends structurally on a densely detailed connectivity and cohesion. So does a scientific experiment, of course, but that is a means to an end: once it has performed its task of verification or falsification it is of no value in itself. Furthermore it must be repeatable to be valid – a crucial element in the plot of Carl Djerassi's first novel, *Cantor's Dilemma* – whereas a work of literature is unique, and cannot (outside the fiction of Borges) be repeated by another writer.

From different parts of Carl Djerassi's memoirs we can string together a fascinating account of how he changed from scientist to creative writer. It is a story that carries faint echoes of John Stuart Mill's famous conversion from the rigid utilitarianism of his father to a broader liberal humanism, triggered by an episode of deep depression from which he recovered with the help of Wordsworth's poetry. Carl Djerassi's transformation began with a more gradual progression from being a "hard" scientist to a "soft" – or at least softer – one, and this in turn was connected with his experience as a teacher. Chemistry itself is a hard science, second only to physics in that respect, he says. But applied chemistry led him to the invention of the Contraceptive Pill, and that in turn involved him in the politics, ethics, and sociology of birth control, a dimension of his research that conventional courses in chemistry did not touch. At Stanford in the late sixties and early seventies – that era of youthful revolution and educational experiment – he devised a postgraduate course on Biosocial Aspects of Birth Control which challenged all pedagogic orthodoxies. Students were organized into teams, and sent out to do original research and report back in any form they liked. Some presented their findings dramatically, and we can see here perhaps the seed of Carl Djerassi's later career as a playwright. The course was a roaring success. One former student wrote to Carl that it "was *the* most important course I have ever taken in my career as a student.... You taught me how to fish instead of simply giving me a fish to eat when I was hungry." (Djerassi 1992, 146)

Another crucial factor in Carl Djerassi's evolution as a writer was his relationship with Diane Middlebrook, partly because she was herself a writer and teacher who widened his knowledge and deepened his enjoyment of literature. He began to read poetry, especially Wallace Stevens, and write poems himself, some of which were published. When they had been lovers for some years, Diane went to the Radcliffe Institute at Harvard for a year's sabbatical and wrote to say she was leaving him for another man. Shocked by this rejection and tormented by jealousy, Carl relieved his feelings first in poetry and then with a novel, called *Middles,* closely based on the experience. Happily Diane and he were reconciled a year later, and married soon after that, but he had caught the fiction-writing bug and would never get it out of his system. Diane made him promise not to publish *Middles,* probably wisely, but he cannibalized parts of it in some of the short stories he began to write and published in a collection called *The Futurist.* One of these stories, however, called "Castor's Dilemma," had its source in his professional, not his emotional life, and he later developed it into a full-length novel under a slightly different title.

Cantor's Dilemma was a deserved success both with literary critics and the scientific community. Some members of the latter complained about "washing dirty lab coats in public," but that's something every campus novelist must put up with. Most recognized its accurate picture of the world of academic science, which the author summarized in an afterword: "Publications, priorities, the order of the authors, the choice of the journal, the collegiality and the brutal competition, academic tenure, grantsmanship, the Nobel Prize, *Schadenfreude* – these are the soul and baggage of contemporary science." (Djerassi 1991, 229f.) The novel has been widely prescribed for reading by science students in colleges and universities. Thus writing that was inspired partly by the experience of teaching, fed back into teaching – and continues to do so. Carl himself took the process further, devising a course, Medicine 256, for postgraduate and post-doc medical students at Stanford, whose members had to write works of fiction exploring some issue in medical ethics, which were then circulated and discussed under conditions of anonymity by the group. He describes it as the most exciting course he ever taught at Stanford. But there may be a more exciting one yet to come. He is still teaching there occasionally and tells me in a recent letter: "I am offering a heavily over-subscribed cross-disciplinary seminar which involves two departments that never before have collaborated: chemistry and drama. The title of the course is 'science-in-theatre: a new genre?'"

Of all the many lives Carl Djerassi has led perhaps the most essential, in the sense that it motivates and energizes most of the others, is that of the teacher. He has an irrepressible, inexhaustible urge to instruct, to convey knowledge, especially scientifically-based knowledge, and most of his creative writing has been fuelled by it. The novels and plays are laced with wit, often of a saucy sexual character, and have intriguing plots, but their basic thrust is uncompromisingly intellectual and educational. Often they have their origins in some problem or phenomenon (not always strictly "scientific" in nature) which has attracted his attention and the work is the product of his own intensive research into the subject. After the success of *Cantor's Dilemma,* he recalls, "I decided on a literary gamble: to pursue fiction writing for didactic reasons" (Djerassi 2001, 164), regarding this project as a gamble because contemporary literary culture is suspicious of didacticism in art. But his work is not really didactic, in the sense of conveying a dogmatic message; it is *dialogic*, and has become increasingly so as he favored drama over prose fiction as a medium. A subject of some importance and complexity is discussed and debated from various points of view, and the audience learns a great deal from listening, but must make its own decisions about the elusive "truth" of the matter, which might be the ethical and inter-personal consequences of new developments in *in vitro* fertilization (*An Immaculate Misconception*), or what defines original discovery in scientific research (*Oxygen*), or the nature of Jewish identity (*Four Jews on Parnassus*). Carl Djerassi has more than once described the liberation he felt in escaping the restraints of professional scientific expository prose, in which the first person pronoun is taboo; and in his preface to *Four Jews on Parnassus* he writes:

In my former incarnation as a scientist over half a century, I was never permitted, nor did I allow myself, to use direct speech in my written discourse. With very rare exceptions, scientists have completely departed from written dialogue since the Renaissance, when, especially in Italy, some of the most important literary texts were written in dialogue – expository or even didactic to conversational or satirical – that attracted both readers and authors. Galileo is a splendid case in point. (Djerassi 2008, xvif.)

Carl Djerassi's fondness for the dialogue as a form of discourse is another reason to regard him as a modern reincarnation of Renaissance man.

Works Cited

Djerassi, Carl. 1991. *Cantor's Dilemma*. New York: Penguin.
Djerassi, Carl. 2008. *Four Jews on Parnassus: A Conversation. Benjamin, Adorno, Scholem, Schönberg*. New York: Columbia UP.
Djerassi, Carl. 1992. *The Pill, Pygmy Chimps, and Degas' Horse*. New York: BasicBooks.
Djerassi, Carl. 2001. *This Man's Pill. Reflections on the 50th Birthday of the Pill*. Oxford: Oxford UP.

Mehr als ein „Renaissance Man":
Die Präsenz von Carl Djerassi

Laudatio zur Verleihung der Ehrendoktorwürde der Technischen Universität Dortmund an Carl Djerassi, 23. April 2009

Hans Ulrich Gumbrecht

Obwohl der Sport, wenigstens Sport aus der Perspektive des faszinierten Zuschauers, zu den wenigen Leidenschaften in der Kultur unserer Zeit gehört, welche Carl Djerassi – bis heute jedenfalls – kalt gelassen haben, will ich meine Lobrede auf ihn mit einer Erinnerung an zwei Sportler (und an einen sportlichen Wissenschaftler) auf den Weg bringen. Es ist ungefähr ein Jahrzehnt her, dass Stanford, Carls und meine Universität, die gute Idee hatte, nach einem hinreißenden Vortrag des damaligen Schachweltmeisters Gary Kasparov (über die Unverzichtbarkeit physischer Fitness im heutigen Leistungsprofil seines Spiels) den Redner zusammen mit Stanfords damaligen Head Coach für American Football, Tyrone Willingham, und mit Terry Winograd, einen Kollegen, der heute als einer der Begründer der Disziplin „Computer Science" gilt, zu einer Podiumsdiskussion über das Thema „Strategien" zusammenzubringen. Willingham sollte die Debatte mit einer Frage an Kasparov eröffnen. Aber als der Moment gekommen war, stand Willingham vom gemeinsamen Tisch der Podiumsdiskutanten auf und sagte langsam, mit bewegter und doch fester Stimme: „Mr. Kasparov, before I ask the first question, I need to tell you that it is an honor to be in presence of such greatness." Ich möchte, lieber Carl, in der feierlichen Öffentlichkeit dieses Nachmittags dieselben Worte an Dich richten, bevor ich über Dich spreche: ich danke Dir dafür, dass Du bei uns bist, denn „es ist eine Ehre, in der Gegenwart solcher Größe zu sein."

Für mich selbst kann ich dies sagen, seit Du mich Neuankömmling im September 1989 zum Mittagessen in den Stanford Faculty Club eingeladen hast, und ich bin sicher, dass es auch für die vielen Studierenden und Kolleg/innen gilt, die sich heute in diesem Raum versammelt haben, um Dich zu ehren. Wir spüren, wie es die Präsenz Deiner Größe bewirkt, dass wir die eigene Existenz mit besonderer Intensität spüren, und zugleich erinnert uns Deine Präsenz an die Potentiale – an die unbegrenzten Potentiale vielleicht – der menschlichen Existenz überhaupt und unserer eigenen Existenzen als Teil von ihr. Auf mich hat die Gegenwart von Größe wie Deiner die manchmal etwas verwirrende Wirkung, dass ich mir Geschichten vorstelle, die in einem philosophischen Sinn, wie ihn Aristoteles in der *Poetik* beschreibt, gewiss wahr sind, ohne dass man sie guten Gewissens der Dimension des Faktischen zuschlagen könnte. In diesem Sinn weiß ich nicht mehr, ob die Erinnerung ein Produkt meiner Fantasie ist oder die faktische Wahrheit, dass Du mich in jenem September 1989 deshalb zum Lunch in den Faculty Club einludst, weil damals die deutsche Übersetzung Deines ersten Romans, *Cantor's Dilemma*, in täglichen Folgen auf den Seiten der *Frankfurter Allgemeinen Zeitung* erschien (das jedenfalls ist ein Faktum) und ein Freund Dir gesteckt hatte, dass der neue deutsche Kollege in Stanford Doktorvater des ebenfalls neuen „Literaturchefs" jener Zeitung sei (auch das ein Faktum); und Du den (logisch eigentlich bestechenden) Schluss gezogen hattest, ich müsste also meinen ehemaligen Doktoranden auf Deinen Roman verwiesen haben. Heute wiederhole ich nur, lieber Carl – allerdings „öffentlicher als je zuvor" – was ich Dir schon so oft gesagt habe: ich kann dieses Verdienst, so gern ich es täte, leider nicht in Anspruch nehmen. Allerdings ist mir nicht entgangen, dass meine wiederholten Geständnisse Dich bisher immer nur noch fester in dem mich ehrenden Irrtum bekräftigt haben.

Faktisch ebenso wie philosophisch wahr ist freilich, dass Dein Name einer von dreißig war, als im späten Dezember 1999 die *London Times* die größten Gestalten des zweiten Millenniums benennen und ehren wollte – was unserem Gefühl, heute von der Präsenz Deiner Größe geehrt zu sein, eine fast atemberaubende geschichtliche Tiefenschärfe gibt. Faktisch und philosophisch wahr ist schließlich auch, meine Damen und Herren, dass Carl Djerassi diese bemerkenswerte, in ihrer Tragweite ja kaum zu überbietende Wahl (nach der pflichtgemäßen Abwicklung einiger Bescheidenheitstopoi) beistimmend kommentiert hat: zumal wenn die Existenz der Anti-Konzeptions-Pille ein weiteres Auseinanderstreben von Sex und menschlicher Reproduktion befördere, sagte er, sei das Ranking unter den wichtigsten Entdeckungen der letzten tausend Jahre gerechtfertigt. Für meinen Teil finde ich kaum Worte, um zu sagen, wie sympathisch mir solche Unverstelltheit ist. Doch viel mehr Kollegen beklagen sich natürlich darüber, dass Carl eingebildet sei („full of himself") oder narzisstisch – und darauf habe ich mir angewöhnt, nicht mehr mit Relativierungen oder gar mit Entschuldigungen für Carl zu antworten (wer will denn außerdem schon nicht narzisstisch sein in unserer akademischen Welt?), sondern mit der Feststellung, dass jemand nach so einer Lebensleistung nicht nur Anlass, sondern alles Recht hat, stolz auf sich selbst zu sein. Vielleicht gibt es ja eine Dimension von Bedeutung, wo Bescheidenheit keine Tugend mehr ist, sondern nur ein Symptom der Unfähigkeit und Verlegenheit im Umgang mit der eigenen Größe.

Das bringt mich zum Thema oder genauer (denn Thema ist natürlich „Carl Djerassi") zur spezifischen Perspektive meiner Lobrede. Ich möchte über Carl Djerassi als „Renaissance Man" sprechen, einmal, weil ich weiß, dass unser Ehren-Gast diesen Begriff gerne hat und nicht selten selbst benutzt (den man ins Deutsche wohl am ehesten mit dem viel neutraler klingenden „Universalgenie" übersetzt). Aber ich wähle diesen Begriff auch und vor allem, weil Carl Djerassi sehr sichtbar die kaum fassbare Vielfalt von Rollen des Universalgenies ist, denen er sich gestellt, die er verkörpert und die er zum Teil sogar selbst erfunden hat (und sicher noch erfinden wird). Sie machen jenen Teil seiner Präsenz aus, den man anerkennen und in der Syntax von Enzyklopädien komprimieren kann. Über den „Renaissance Man" hinaus aber gibt es einen Carl Djerassi, von dem der öffentliche Carl Djerassi nicht redet und nur in Andeutungen schreibt. Auch die Präsenz dieses fast unsichtbaren Carl Djerassi spüren wir alle in diesem Moment, obwohl es naturgemäß schwerer ist, sie in Gedanken und Begriffen zu erfassen; als seine private Persönlichkeit und Geschichte ist sie Bedingung der Möglichkeit für den offiziellen Carl Djerassi – und dies ist Grund genug, heute auch von ihr, so gut es geht, zu reden.

Die Beschreibung des offiziellen Carl Djerassi muss mit dem Chemiker beginnen, denn diese Rolle hat ihn zuerst weltweit berühmt gemacht. Der Chemiker Djerassi hat wohl alle Preise gewonnen, welche das Fach und der Beruf international bereithalten, außer dem Nobel-Preis, mit dem nicht wenige seiner Schüler geehrt worden sind (und ich frage mich wieder, ob es eine meiner philosophisch korrekten Imaginationen ist, zu denken, dass ihn der Nobel-Preis bisher umgehen musste, weil er einer der Allerhöchst-Privilegierten ist, die der Schwedischen Akademie Kandidaten für diese Ehrung vorschlagen). Faktisch richtig ist jedenfalls, dass Carl Djerassi zu jenen Naturwissenschaftlern des zwanzigsten Jahrhunderts gehört, deren Arbeit am nachhaltigsten zur Veränderung der *conditio humana* und ihrer biologischen Umwelt beigetragen hat: durch die führende Rolle, die er bei Erfindung der Anti-Konzeptions-Pille gespielt hat; durch seine Beiträge zur Entwicklung des Cortisons; und durch seine – seltener genannten, aber möglicherweise ebenso lebens-wichtigen – Leistungen bei der Produktion von Insektenvernichtungsmitteln. Die Gilde der Naturwissenschaftler vor allem sollte, über alle denkbare Rivalität hinaus, dem Finanzgenie Carl Djerassi dankbar sein. Denn er war einer der ersten unter ihnen (wenn nicht der erste überhaupt), der seine zukunftsträchtige Erfindung nicht für eine möglichst hohe Summe verkaufte, sondern sich von einer kleinen Firma in Mexiko, welche die Pille zuerst produzierte, in Aktien bezahlen ließ (damit hat er die Geschäftspraxis von „Microsoft" und anderen Silicon Valley-Firmen um ein knappes halbes Jahrhundert vorweggenommen). Natürlich hat Carl Djerassi dieses Aktienpaket eben genau zur rechten Zeit wieder verkauft und überhaupt zeit seines Lebens bei vielfachen finanziellen Einsätzen von Intuitionen profitiert, um die ihn Finanz-Profis beneiden. Ich weiß auch, dass Du Dich kaum freuen wirst, mein lieber Carl, diese Facette hier erwähnt zu hören (sie kommt Dir wohl zu handwerklich vor?) – aber ohne sie wären Dir einige Dimensionen verschlossen geblieben, die zu einem echten „Renaissance Man" gehören. Etwa der schrittweise-geduldige Erwerb Deiner Sammlung von Werken aus der Hand Paul Klees, die heute zu den bedeutendsten Klee-Sammlungen weltweit gerechnet wird, weil sich in ihr ein besonderer Blick und eine besondere Stimmung gefunden und artikuliert haben. Oder die Gründung und ständige Betreuung des Djerassi Resident Artists Program, dank dessen es in den vergangenen Jahrzehnten mehr als tausend jungen Künstlern und Künstlerinnen möglich wurde, über mehrere Monate in der erhabenen Landschaft eines Areals zwischen der Bay von San Francisco und dem Pazifik ihre Arbeit weiterzuentwickeln.

Schließlich gibt es den literarischen Autor Carl Djerassi, also jene Rolle, welche ausschlaggebend für die heutige Ehrung durch die Technische Universität Dortmund war. Ich selbst bewundere nicht zuletzt einige Deiner Gedichte – aber vielleicht hat man ihnen noch nicht die verdiente Aufmerksamkeit gewidmet, weil es Dir ja gelungen ist, im Roman und im Drama einen ganz neuen Diskurs (oder zwei neue Gattungen) durchzusetzen, den Du selbst „Science-in-Fiction" bzw. „Science-in-Theater" nennst: der Gebrauch des Repertoires literarischer Formen und Figuren zur Vergegenwärtigung der Welt und der Probleme der Naturwissenschaften. Mir haben Deine Erzählungen diese Welt – vergleichsweise spät im Leben – mit der Intensität einer frisch geweckten Neugierde so erschlossen, dass ich sie vor allem nicht mit einem „didaktischen" oder „pädagogischen" Talent auf Deiner Seite assoziieren möchte. Denn das käme mir zu klein vor. Eher spüre ich in Deinen literarischen Texten, Kapiteln und Sätzen oft den Atem und die Energie eines Genies, das sich nicht auf eine einzige Welt oder Dimension beschränken lässt.

Und auch deshalb, nicht nur wegen der Vielfalt der Rollen, sondern wegen der einen Energie, bist Du im prägnanten historischen Sinn ein „Renaissance Man". Was jene großen Charaktere der frühesten Neuzeit in den italienischen Stadtstaaten anders machte als die Helden und Heiligen des Mittelalters, war, dass sie alle ihre einzelnen Rollen von außen sahen und dass sie sich auf keine von ihnen festlegen ließen, weil ihnen das am Ende ermöglichte, Beobachter einer Welt zu sein, die ihnen als Material der Gestaltung zur Verfügung stand. Es ist mithin keine Bildungs-beschwerte Festtags-Rhetorik, lieber Carl, wenn ich gestehe, dass Du mich oft an den hohen Mut dieser Gestalten erinnert hast – und auch an die Gefahr der Hybris, die an ihr Welt-Formen (ja sogar noch an meine Wahrnehmung davon) unvermeidlich gebunden ist. Deshalb, um zu unterstreichen, wie ernsthaft ich gerade an dieser Stelle bin, will ich Deinen Namen auch nicht neben einzelne

Namen jener Renaissance-Figuren stellen (denn das müsste ja fast wirken wie die Benutzung von Abziehbildern) – sondern stattdessen eine Fähigkeit betonen, welche Machiavelli herausstellte, als er sich fragte, warum Fernando von Aragón der einzige Fürst der eigenen Zeit war, in dem er bereits sein Idealbild vom „Fürsten" verwirklicht sah. Es war das Schnelle, antwortete er, die Behändigkeit in Fernandos Gedanken und seiner Fähigkeit, Rollen zu wechseln, welche die Zeitgenossen in Bann schlug und überwältigte. Sie ähneln der schönen Mühelosigkeit und Energie, mit der Du auch in Deinem neunten Lebensjahrzehnt zwischen Kontinenten und Kontexten, Rollen und Tonalitäten so schnell changierst, dass am Ende nur die anscheinend keine Grenze kennende Energie und Sicherheit des Wandels permanent mit Dir assoziiert werden kann. Ich vermute, dass dies eine Energie und eine Sicherheit sind, die Deiner eigenen, ebenso genauen wie rationalen Planung schon immer vorausgehen.

Energie und schöne Unruhe artikulieren sich natürlich nicht allein in den Dimensionen und Gesichtern des öffentlichen „Renaissance Man" Carl Djerassi. Zugleich hinterlassen sie Spuren, die vor diesem Hintergrund beinahe „dysfunktional" aussehen mögen – und uns doch näher zu den Quellen der Energie bringen, welche Gedanken und Taten beseelt. Viele dieser scheinbar dysfunktionalen Spuren verweisen auf Gesten und Schichten, wegen derer ich Carl Djerassi – über all die Bewunderung hinaus – wirklich liebe. Zum Beispiel habe ich noch niemanden gesehen, der mit seinen finanziellen Möglichkeiten großzügiger umgeht als er, und dieses Urteil hat einen sehr prägnanten Grund. Ich habe nämlich die Gewissheit, eigentlich seit unserem allererten Gespräch, dass Carls Großzügigkeit ausnahmslos den Sinn hat, den Beschenkten unabhängiger und autonomer zu machen. Die Djerassi Art Foundation zum Beispiel will vor allem Künstlern auf der Schwelle zum Durchbruch weiter helfen – widmet sich also weder den noch kaum geformten Lehrlingen, noch den vollendeten großen Meistern. Genau in diesem Sinn kann ich mich wirklich an keine Begegnung mit Carl erinnern, aus der ich nicht eine Ermutigung mitgenommen hätte und zugleich eine neue Aufgabe, die auf die Entdeckung des eigenen Potentials gerichtet war. Weil er in dieser besonderen Weise großzügig ist, konnte es Carl Djerassi auch gar nicht vermeiden, politisch – zumal unter amerikanischen Bedingungen – zur Linken zu gehören. Sein Protest gegen die militärische Intervention der Vereinigten Staaten in Vietnam war so unmissverständlich, dass sie ihn auf eine offizielle Liste von „Feinden Amerikas" brachte. Sozialpolitische Initiativen im Land haben immer mit seiner Unterstützung und seinem Rat rechnen können. Zu diesen Spuren gehört auch die Erfahrung, dass sich Carl Djerassi, der ja sehr wohl weiß, wer er ist, kaum um institutionelle Hierarchien und seinen eigenen Rang in ihnen kümmert. Als Carl meinem Department in Stanford vor Jahren sein elegantes Haus in den Bergen zwischen der Bay von San Francisco und dem Pazifik Ufer für einen Tag intensiver Reflexion zur Verfügung stellte, schleppte er mehr Stühle und mehr Pakete mit Getränken für die Teilnehmer des Gesprächs als irgendeiner meiner Kollegen – und das war wohl ganz einfach so, weil er sich es anders nicht vorstellen kann. Als wir gemeinsam die großen Künstler Christo und Jeanne Claude eingeladen hatten, bestand er darauf, sie persönlich am Flughafen von San Francisco abzuholen und wartete dann viele Stunden auf einen sich immer weiter verspätenden Flug, weil es ihm wichtig war, seine Bewunderung für ihre Arbeit vom ersten Moment der Begegnung an physisch deutlich zu machen.

Niemandem bin ich aber auch begegnet, der so gnadenlos hart mit sich sein kann wie Carl Djerassi. Er hat immer wieder – und ohne alle Koketterie – davon gesprochen und geschrieben, wie schwer es sein muss, in seiner Nähe oder gar als sein Verwandter zu leben. Er hat es sich auferlegt zu beschreiben, wie seine Tochter Pamela 1978 ihrem Leben ein Ende setzte – in eben jenem paradiesischen Gelände, das heute die Djerassi Art Foundation aufgenommen hat. Von dieser erschreckenden und im Sinn der griechischen Tradition wirklich „tragischen" Verantwortung hat er sich nicht entlastet, obwohl ihm und der Familie bekannt war, dass Pamela mit einer endogenen Depression kämpfte. Carl Djerassi, der „Renaissance Man" der unendlich vielen Rollen, ist auch ein Mann der Treue. Ich habe ihn nie so konzentriert gesehen, so in sich und auf die Vergangenheit gekehrt wie bei der Gedenkfeier unter Freunden für seine Frau Diane Middlebrook, als er denselben Anzug trug wie bei Dianes und seiner Hochzeit zwanzig Jahre zuvor. Diese Tode waren Situationen, wo es Carl nicht gelungen war, seine Intelligenz und auch sein Vermögen einzusetzen, um die Grenzen zwischen „unmöglich" und „möglich" zu verschieben. Wäre eine medizinische Innovation in Reichweite gewesen, um Dianes Leben zu erhalten, dann hätte Carl seine Frau nicht nur täglich gepflegt, sondern auch alles daran gesetzt, die entfernteste Möglichkeit des Unmöglichen wirklich werden zu lassen. So wie er sich selbst vor einem Vierteljahrhundert – noch vor seiner Ehe mit Diane – nicht der ihm gestellten Diagnose einer unheilbaren und schnell zum Tod führenden Krebs-Erkrankung unterwarf; so wie er zu einem eindrucksvollen literarischen Autor in einer Sprache wurde, die nicht die Sprache seiner Kindheit war; so wie er der einzige ist, der überhaupt nicht peinlich wirkt, wenn er sein hohes Alter zur Jugend erklärt. Denn keinen habe ich gesehen, der mit soviel Energie und Geschick das Leben für die Möglichkeiten dessen offen hält, was als unmöglich gilt.

Woher aber kommt diese Energie, die nicht vorsichtig unterscheiden lässt zwischen jenen vielfachen Manifestationen, die einen „Renaissance Man" ausmachen, und den Spuren einer vielleicht noch eindrucksvolleren Größe, die nie öffentlich wird? Natürlich gibt es auf Fragen dieser Art – glücklicherweise – weder historische noch genetische Antworten, die uns überzeugen könnten. Es ist also ein archaischer, ein mythologischer eher als ein rational-wissenschaftlicher

Impuls, der uns auf Carl Djerassis Familie verweist. Sein Vater kam aus einer sephardischen Familie, die seit der Vertreibung aus Spanien im späten fünfzehnten Jahrhundert im heutigen Bulgarien gelebt hatte. Die Mutter gehörte einer aschkenasischen Familie in Wien an, die entfernt mit der Familie Sigmund Freuds verwandt war. Carl Djerassi ging auf dieselbe Realschule in Wien, die auch Freud besucht hatte, und sein Vater hatte als Medizinstudent noch Vorlesungen des Privatdozenten Freud gehört. Die Eltern lernten sich beim Medizinstudium kennen, obwohl ihre Wege der Spezialisierung denkbar weit divergierten: der Vater wurde Spezialist für Geschlechtskrankheiten und die Mutter, etwas älter als der Vater und den Fotografien zufolge eine Schönheit, wurde Zahnärztin. 1923 kam Carl zur Welt und wuchs bis 1929 in Sofia auf. In jenem Jahr ließen sich die Eltern scheiden, Carl ging mit der Mutter nach Wien und verbrachte die Ferien regelmäßig in Sofia. Die Eltern waren rücksichtsvoll und zärtlich genug, um in Gegenwart ihres Sohnes über die Scheidung nie zu sprechen. 1938, auf der Flucht vor den Truppen des nationalsozialistischen Deutschlands, die Österreich annektierten, entkamen Carl Djerassi und seine Mutter nach Sofia, wo die Eltern sich wieder verheirateten, um Sohn und Mutter die legale Auswanderung aus Bulgarien zu ermöglichen.

In dieser Vergangenheit berührt sich Carl Djerassis Leben spannungsvoll mit der Vorgeschichte meines Lebens – und gewiss, meine Damen und Herren, mit den Familiengeschichten von vielen unter Ihnen. Mein Vater hätte seinem Alter und auch seiner Einstellung nach durchaus einer jener deutschen Soldaten sein können, die ins vielerorts begeisterte Österreich einmarschierten, und ich weiß, dass der Vater meiner Mutter ausgerechnet hier in Dortmund das Ereignis der Annexion mit vielen Sektflaschen feierte. Carl aber gehört zu jenen Überlebenden der deutschen Bedrohung im Zweiten Weltkrieg, die mit fast extremer historischer Versachlichung den Schrecken auf Distanz gesetzt haben. Wahrscheinlich findet er die Erinnerung an meinen Vater und meinen Großvater allzu weit hergeholt in einer Rede über sein Leben. Etwas hilflos gestehe ich dann ein, dass – trotz Carls Entscheidung zur Nicht-Befassung – das Problem dieser historischen Kontiguität mein Problem bleibt und, meine ich, unser Problem, das Problem derer, die damals in Deutschland geboren sind.

Carl findet es, weiß ich, interessanter, dass er mit fünfundzwanzig Jahren, 1948, im Jahr meiner Geburt, schon in Mexiko Stadt an jenen chemischen Problemen arbeitete, deren Lösung zur Pille führen sollte. Er wird aber wahrscheinlich – wie ich es war – beeindruckt sein zu hören, dass unsere Mütter für uns als kleine Jungen dasselbe Abendgebet aufsagten, mit dem man sich an den jüdischen und an den christlichen Gott richten kann: „Müde bin ich, geh zur Ruh, / schließe meine Augen zu. / Vater lass die Augen Dein / über meinem Bette sein." Wir beide haben einen kaum zu stillenden Ehrgeiz, als Autoren im amerikanischen Englisch anerkannt zu werden – aber müssen mit einigem Schmerz zugeben, dass wir unseren deutschen Akzent im gesprochenen Englisch selbst kaum hören (und schon gar nicht mit bleibendem Erfolg korrigieren) können. Beide sind wir Amerikaner geworden, um der deutschen Geschichte zu entkommen. Carl musste ihr entkommen, um physisch zu überleben. Ich wollte ihr entkommen, indem ich Deutschland verließ, damit es mir – trotz der Scham über vieles, was Verwandte der vorigen Generationen getan hatten – möglich wäre, diese Geschichte, als einzige, die ich habe, in der Distanz eines anderen Landes an meine Kinder weiterzugeben.

Dass ich nun ausgerechnet in Dortmund Carl Djerassi im Namen einer deutschen Universität sagen darf, wie sehr uns die Gegenwart seiner Größe ehrt, fühlt sich an wie das Geschenk eines Gottes, an den ich nicht glauben kann. In Wirklichkeit ist es natürlich – für uns alle – das Geschenk der Größe, der Großzügigkeit und der Kühnheit von Carl Djerassi, hier und heute mit seinem Werk und seinem Leben, die sich nicht trennen lassen, das Unmögliche ermöglicht zu haben. So many thanks for that, dear Carl.

The Two/Several Cultures

Cross Talk:
A Dialog between Carl Djerassi and George Klein

Carl Djerassi and George Klein

30 June 2008
George Klein to Walter Grünzweig

Many thanks for your letter of June 28. I am honoured by your kind invitation to participate in the symposium on the occasion of Carl Djerassi's 85th birthday. ...

Carl and I are old friends since many years and I admire him for his many contributions, within and outside science. Still, our approach to write about science and scientists is quite different and while I am very impressed by Carl's writings, I like to see science and scientists in a very different light.

It strikes me that a dialogue between Carl and me, if he would also feel like it, might be the best approach to my possible contribution to the meeting. We would agree on many points, perhaps disagree on even more, and we might conclude by agreeing to disagree.

Would you explore with Carl whether he would favour the idea?

22 September 2008
George Klein to Walter Grünzweig

As for title and abstract, however, there is a problem. As we have agreed before, I prefer a dialogue to a lecture. You have then indicated that a dialogue may be possible, although I do not have the details. Does this mean a dialogue with Carl himself? That would be, of course, most appropriate. I imagine that we could discuss the ways in which science can be communicated through literature and theatre. Carl and I approach this quite differently and, in many respects, he succeeds better than I. He tends to communicate more on the level of the public, whereas I prefer to keep a certain distance and remain more "serious." One could also say that Carl hopes to get across the message through a more populistic approach, whereas I tend to try to raise the public interest for the ways scientists think and work.

6 October 2008
George Klein to Walter Grünzweig

As a title, I would suggest: "Bridging the gap between science and literature – a dialogue between Carl Djerassi and George Klein." (Of course, Carl should be allowed to modify this in any way!) An abstract (very artificial!) looks as follows:

The divide between CP Snow's two cultures is still with us. Carl Djerassi's novels and plays represent one attempt to bridge the gap and bring science or, rather, the world of the scientists, closer to the lay reader. George Klein's writings have a similar purpose, but a different approach. The two approaches will be contrasted against each other in the hope that, while divergent, they may be mutually complementary.

8 January 2009
George Klein to Carl Djerassi

I am glad that you have agreed to a dialogue. It will have to be, and it will be, spontaneous, but at the same time it is important that we know a little bit in advance what we can talk about and where we stand. Do you agree?

Several aspects of your literary oeuvre appeal to me and some of them I admire. But I am also critical – or perhaps dissatisfied is the right word – about some other aspects. The question is whether we should focus on the former, the latter, or both.

I shall try to summarize my views.

On the positive side, I appreciate your inventiveness, your style, your sense of humour, the boldness of your conjectures and the complexities of your intrigue. So, what is my criticism?

You convey to the public – as Jim Watson did before – that scientists are all too human, with all the frailties, weaknesses and passions. Their jealousies and hatreds are no different, their egocentricity is not less and sometimes more that of average mortals. But you don't really show, at least to my feeling, the other side of the coin.

How shall I put it? Perhaps by saying this: In their subjectivities, scientists are like other people. Their ambitions and frustrations, their sex drive, their greed, their will power, their manipulations and intrigues are not more interesting than what you find in other people. What makes them interesting is their scientific quest, their striving and, if you speak of the best scientists, their relentless search for understanding an ever so little segment of the world around us and inside us. I may not have read enough of your writings, and I may be wrong, but from what

I have read, I don't have the feeling that you are really showing this to your lay readers.

Perhaps you depart from the notion that the public would not understand how scientists work and it is better to present scientists from their generally human side. If that would be the case, it would be an underestimation of the intelligent reader.

Jim Watson, while dwelling a lot on the more superficial aspects of scientific lives, could nevertheless manage to put through the content, the excitement of science itself. In spite of all the gossip, it reigns supreme in "The double helix" and, to a lesser extent but still quite clearly, in his latest book "Avoid boring people." In between those two, he has some more gossipy and less interesting books.

The two scientific autobiographies I would single out as superb in conveying the essence of the scientific endeavour are François Jacob's *La statue intérieure* and Sydney Brenner's *My life in science*.

Dear Carl, I hope you don't mind my being frank to you, but how could we otherwise have a dialogue? I am quite open to your counterarguments and I really look forward to our discussion.

9 January 2009
Carl Djerassi to George Klein

Dear George,

I am exceedingly pleased with the fact that you are willing to have the kind of dialogue you mention and I welcome your desire to address primarily (though hopefully not exclusively) the critical aspects of my writing you outlined.

There is only one caveat. I need to know on which of my many fiction and play books and 3 (!!!!) autobiographies you base your conclusions. Otherwise, it may just turn out to be a discussion of what I have written in a couple of books and the response from me would then be simply, "yes, but look at what I have written in . . . ."

It most certainly is not necessary for you to have read all of my books, but in terms of the criticism you raise (and which we definitely should discuss, because it is important), it is really necessary that you must have read or at least skimmed the following:

a) my three autobiographies:
    Steroids Made it Possible. by Carl Djerassi (Scientific Autobiography) American Chemical Society Books, Washington, DC; 205 pages (1990).
    The Pill, Pygmy Chimps, and Degas' Horse: An Autobiography. by Carl Djerassi. Basic Books, New York, 319 pp. (1992).

This Man's Pill – Reflections on the 50th Birthday of the Pill, by Carl Djerassi (paperback) Oxford University Press, London, New York, 2003, 308 pp.

b) the following three novels of which the second, in particular, addresses the criticism you expressed. (Incidentally, I also thought that Jacob's autobiography was first class.)
Cantor's Dilemma. by Carl Djerassi (Paperback Edition) Penguin Books, New York, 230 pages (1991).
The Bourbaki Gambit, by Carl Djerassi (paperback edition), Penguin-USA, New York, 230 pp. (1996).
NO. by Carl Djerassi (Novel – Paperback Edition) Penguin Books, New York, 276 pages (2000).

c) At the very least, one play, namely "OXYGEN" (Wiley, VCH, Weinheim 2001) because it contains many of the "romantic" views expressed by you (which are also shared by the co-author Roald Hoffmann), whereas I contributed more of the "critical" ones.

All of these books can be gotten new or used from *Amazon* which would probably be the simplest way of getting them in Sweden unless your university library has them. Quite separate from these books, you might be interested on Jewish or literary grounds (but not scientific ones) in my newest book (FOUR JEWS ON PARNASSUS). I am sending you a poster image for a reading I am giving here next week.

11 January 2009
George Klein to Carl Djerassi

(to which Carl Djerassi replied the underlined on the same day interspersed between George Klein's sentences)

I am glad that we can have this correspondence prior to our dialogue. I feel that there is still some misunderstanding.

Which books have I read? Your autobiography, *The Pill....*, *Cantor's Dilemma*, *Oxygen*, I have read from cover to cover. Several of the others I have glanced through fairly superficially. But I don't think that this is the issue. It is not a question of one story or another, or what has been written here or there. I shall try to define what I feel is quite a fundamental difference in our view on science and scientists or, more precisely, on the aspects of science we try to convey to non-scientists.

Referring to Roald Hoffmann in your letter as sharing my view, you speak about this as a "romantic view." This immediately tells me that you don't really understand what I am trying to say.

Carl, we are originally both Central European Jews, from well to do bourgeois families. We both got away and "made it" in one way or another. I know the sources of our drive, what has made us tick, and what keeps us ticking at what others – but certainly not we – would consider as an advanced age.

Both Eva and I *loved* your autobiography. It is honest, it makes a fascinating story, and it describes exactly *that*.

What? The most profound of the drives we received from our ancestry: *the quest for success.*

Already before I started school at the age of six, I knew that for a Jewish boy there are only two choices: to become very outstanding or to end up in the gutter. There was absolutely nothing in between. How did I get that notion? I must have gotten it with my mother's milk (even though my mother was a very modest self-effacing person).

Be that as it may, I was able to sustain, over a long life, a kind of double ego. One was running after success, just as I have been conditioned to do, and the other, a little skeptic somewhere in the back of my brain, never tiring of making cynical remarks, boiling down to the question: what the hell are you doing?

In one of my Swedish books published a few years ago, I wrote, in a totally different context, the following:

"Success mentality has penetrated everything. Like melting wax candle, it has been dripping down on the white pages of our souls. It has infiltrated our pores and solidified. Our skin has become fragile, it is shattered by the smallest 'unsuccess.' It has polluted the air that we are breathing. It came so slowly, it was sneaking through so gradually that we have not immediately noticed how much more difficult it has become to breathe. Only when the blue sky of our childhood started to be perturbed by a brown smog have some of us reacted, but it was too late...

Where success mentality is reigning supreme, there success mentality is reigning. You can feel it in the handshakes, you can see it in the smiles. It stains all questions and answers, it surfaces in most associations and dissociations. It infiltrates friendship and love. At its worst, it can block the way for all other thoughts....

It happens sometimes that I send a text or a book I have written to somebody whose comments and criticisms I am eager to hear. But instead of that, he or she sometimes simply congratulates me to my 'success.' I always experience the empty, conventional phrase as a disappointment and sometimes as an insult."
(From: Georg Klein: *Skapelsens fullkomlighet och livets tragik*, Bonniers, 2005).

How does this apply to our dialogue?

Of course, scientists want to be successful, *but that is not interesting, just as their sex life – another, and quite supreme expression of ambition, or greed – is not interesting.* It is – yes, that is the only correct word – *trivial*.

Let me insert here another metaphor. I consider both Bunuel and Bergman as among the greatest film makers of the past century. But I love Bergman and I dislike Bunuel. Bunuel says: "Look, we are all shit." Bergman says: "Look, we are all shit, but see what heavenly aspirations we have!"

We, scientists, have all the earthly ambition for success. But our best hours are when the wish to succeed or, indeed, the wish to *please* suddenly melts away and we are lifted up by our work or by our discourse to perceive a new and different vista.

This is not romanticism. I have seen it happening in myself, in others, and sometimes at meetings, where it could affect a group of people. When the latter happens it can look like this:

There is a small conference, with motivated people who work in the same broad area, doing different things that interconnect. They are all deeply committed to what they are doing. A pleasant place, away from the large cities and their maddening crowd. It starts out like other meetings: Keen prestige and priority watching, competitive tensions threatening to explode any moment, mutual suspicion, speakers' wish to convey their success, to tell everybody that others have no chance. This is the usual setting. But then something may happen. It happens infrequently, I must admit, but it does happen.

<u>Here is the crux of our "disagreement" – if in fact there is one. What you describe is quite correct and of course has also happened to me. In my autobiography I describe such an event at a Gordon conference in connection with cortisone. More importantly, I describe this in some detail in *The Boubaki Gambit*, a novel of mine that you should read before our debate. But the key point is the rarity of such events and I don't write about the exceptions but the reverse. But that is a debate well worth having and I am looking forward to it.</u>

Everybody realizes suddenly that the scientific problem is most exciting. There has been great progress but there is very much more to do. If conditions are really favourable, *the whole atmosphere changes.* Within a very short time – it can happen between breakfast and lunch or even less time – people open up, tensions relax, competition and even the usual forms of covert or more or less open bragging lose their significance. Everybody suddenly starts seeing a new world, as yet covered by fog, but clarifying. People are delighted, they laugh, they joke, they smile at each other. Old foes are asking interested and even personal questions at lunch.

I have seen it happen. I have also been at meetings where all the conditions were right, but it did not happen.

It happened quite often at the yearly conferences we used to have in Ein Gedi, Israel. Part of it was due to the place. The young Israelis were also important. But the guiding spirit was the scientific problem itself. We all felt that there was a job to do and we were part of it.

It is this excitement, and, indeed, euphoria, that Jacob, or Watson, or Brenner have conveyed so beautifully in their books.

My perception of your texts, Carl, is different. They are about the quest for success. They are very well written, they are often very witty, but that is all they are about and with a vengeance.

You have not read enough of my works to make that generalization.

Maybe I have become hypersensitive to this issue of success mindedness. It is possible that I am turning against my own burden that I have received with mother's milk. Quite possible. But I feel that science and scientists are running an ever increasing danger of becoming the victims of their own success. With ever increasing possibilities, increased prestige, the influence of the industry – about which you know much more than I – the thinking of scientists, particularly young scientists, is transformed. We are continuously at the risk of throwing out the baby with the bathwater.

You write, Carl, that you hope I will not only be critical. But Carl, the whole meeting is to honour you, everybody is going to say nice and good things about you, and I will also say some, but I see my task, also because you chose me and agreed to be my dialogue partner, to be the advocate, not of the devil, not of the romantic either, but somebody who is very concerned where it is all going.

A final point: Your specific topic of artificial fertilization. I think you have written somewhere that, in the future, sex will still be done in bed, for pleasure, but reproduction will be done under the microscope. I am sure that this titillates the interest of many. People are intrigued and fascinated by sex as you know better than I. But I perceive it as a freak that will never have any major impact on one of our greatest blessings and greatest curses, the sex drive. We could discuss that as well, if you like.

That, my dear George, is nonsense. I consider the separation of sex and reproduction one of the giant social problems facing humanity. I lecture about this all the time and cover it in various genres and from various sides. My 2001 memoir *This Man's Pill: Reflections on the 50th Anniversary of the Pill* (Oxford University Press, 2001) and the play *Taboos* in my most recent book *Sex in an Age of Technological Reproduction* University of Wisconsin Press, 2008) deal with that.

By all means let's discuss this subject in Dortmund.

Now that you have read all this, may be you are not keen on having me as a dialogue partner!

I am very keen.

If that is the case and you feel that I may spoil the feast, don't hesitate to say so. I will not be offended the slightest and it will not affect our friendship. There is still time for me to gracefully withdraw.

14 January 2009
George Klein to Carl Djerassi

(to which Carl Djerassi replied the underlined text on the same day interspersed between George Klein's sentences)

Dear Carl,
Thank you for your prompt response. I am not happy about it.

Carl, please realize that I am 83 years old, running a laboratory full time, writing grant applications and papers, and I also try to do my own non-scientific writing. I just don't have the time to do a lot of reading of your books, to be quite frank. If the purpose of our dialogue is to examine me on what I have read and what I have not read we can just cancel the event.

Of course not. I don't want you to read any more books of mine. But you present a very black and white view, assigning to me a totally different view from yours without any nuances. I simply mean that I do not think, write or behave that way. I think it will be a strong debate – after all that is what debates are for, but you cannot ignore some of the many other views about scientific behavior I have expressed in my writing. You are totally entitled to express yours – and I trust you will do so forcefully – but not by assigning a simplistic opposite position to me without then also looking at the evidence. That's really all I meant.

Still, if you can be tempted, I would be happy to send you a copy of THE BOURBAKI GAMBIT. You might find it personally interesting since I focus entirely in it on *working* scientists past the age of 60.

There is a difference of opinion between us on what science is first and foremost about and how it should be presented to the lay reader. That is a fundamental disagreement, we cannot deny it. You say that events where competition is relaxed and the excitement about the scientific problem takes over is an exception, a rarity. I have the opposite opinion. Moreover, I think that it is only worth doing science for those events. I don't think it is worth doing science for competition, success, technical applications, and all that. That is a disagreement that we cannot come around.

Sex and reproduction is your field, not mine. I have no doubt that you know much more about it. I still don't believe that people will reproduce depending on the social problems of the world. They will continue to reproduce because of the sex drive that we share with the rest of the biological world, and nothing else.

That will be a productive and even amusing discussion.

Frankly, I am becoming more and more doubtful about our dialogue. You will assert your views and press me on my reading. I will disagree with you and confess my lack of reading. Where does that take us?

P.S. If I would believe that science should be done mainly because of its applications, and not as a quest for understanding the world outside and within us, and that it is and should be guided mainly by competition, I would stop doing it immediately. Fortunately, I don't.

5 February 2009
George Klein to Carl Djerassi

I am half way through *The Bourbaki Gambit* and will finish it before this letter goes out.

First, I must say that it is very well written. I am not surprised that it has become a bestseller. However, your declared goal to bring science to the people by "science in fiction" is still not met, at least not to my perception of what is important in science.

Unlike *Cantor's Dilemma* and *Oxygen*, competition, personal motives and trivial intrigue is not in equally strong focus, even though it is there as a background. What dominates instead is the monkey play of high society, meticulous attention to style, in relation to such trivia (forgive me) as eating and drinking, the (no doubt often well justified) rancor of professors, and particularly professors approaching retirement age, towards the university administration, and a verbal one upmanship game between male and female professors in the ever lasting – but trivial – battle of the sexes.

Who are my idols? Archimedes, of course. Francis Crick who reads and corrects papers in his bed, literarally to his dying day. Or Barbara McClintock, to whom the fact she is a female of our species and everything that has to do with science as a carrier is of no importance. Or Seymour Benzer to whom eating exotic dishes is of great interest, but not as a status symbol, but as a matter of curiosity, and who continuously "reinvents" himself, not with gambits, but by entering new, unexplored areas, particularly in behavioural genetics and who works until the early hours of the morning, even on days when the fatal brain hemorrhage strikes him.

Nothing could be further away from to the people I mentioned – and I know many others like them – than posturing or role playing. They have not chosen science in order to attain a high position or prestige in society, they are not coveting honors and prizes (McClintock suffers from them so much that she calls them martyrdom) and they could not care less about academy membership and prize awarding ceremonies.

[Addition by Djerassi on 18 September 2011:] McClintock, if she really "suffered" from rewards, is so bizarrely unique that you surely do not expect me to make any generalizations based on her behavior. But why did she list so many awards in her own autobiography for the Nobel Prize? Or take you? Look at on the web where you list (quite appropriately) dozens of your honors (http://www.philosopedia.org/index.php/George_Klein). Why do so, if, as you claim, they are so unimportant?

They do science because they are passionately interested *in the thing itself*, a small segment of the world within or around us. This is their all encompassing interest. Everything else is secondary.

I wrote you previously about your very good book about *Four Jews on Parnassus*. I had a similar criticism. Well written, one learns a lot about these great people. But you don't learn what makes them tick, why they are great.

Carl, you are a great scientist and you are also a very skillful writer, inventive and entertaining. But then, if I cannot agree that your "science in fiction" brings science home to the layman, not science at its best, and possibly at its worst, then should we change the topic? I have originally suggested the topic "writing about science for the layman" but perhaps we could choose another one. How about discussing the "two cultures" and ways how to bridge the gap between the sciences and the humanities? In addition to that, we can discuss the future of reproductive physiology, where you are the expert and my contribution would be that of the interested – perhaps puzzled – layman.

If no one else on the panel does this, we might also touch on the way the Pill has influenced sexual behaviour, what the impact of HIV/AIDS has been subsequently, or any other aspects of the true revolution that your chemical experimentation has lead to.

6 February 2009
Carl Djerassi to George Klein

The basis of our apparent disagreement – though very puzzling to me – is really the following. You cite some glorious exceptions, all of whom I admire, but apparently believe that these 5% of top researchers (or maybe 7.5%) are the norm and that I must feel obligated to write about them. You refer to them as your idols. So why don't you write about them? You are a wonderful writer yourself and have written several books. Is any author obligated to write only on the selected favorite topic of a reader?

If what I write is incorrect, then of course you have a point. But if you really think that I write about tribal characteristics that don't exist among the vast majority of scientists, then you live under a wonderful romantic illusion from which I most certainly do not want to dissuade you. In fact, you are to be envied. You refer to most of the behavioral patterns I describe as "trivial." First of all, "trivial" is a very subjective term, but even if they are trivial, the sum total of so many trivial patterns do affect the whole. This is true of most of human behavior and of must humans, and I choose to write about humans and not idols – imaginary or real ones. As I have stated on several occasions, I have chosen to wash dirty lab coats in public, whereas you want me to focus on the impeccably clean lab coats. But realistically speaking, how many working scientists manage not to dirty their lab coats?

I have no problem whatsoever to discuss or debate any or all of these issues at Dortmund on whatever basis you wish, but I do agree that it might start the symposium on a contentious note. I leave it up to our host whether he wants to start the symposium on that note, place it in the middle, or end with it. I am willing to accommodate his and your preferences and am perfectly happy to let you two decide.

7 February 2009
George Klein to Carl Djerassi

Many thanks for your letter of February 6. Here are some responses, comments and suggestions.

You speak about the names I mentioned with admiration as "glorious exceptions." You ask why I do not write about them.

I have written about them, quite a lot. Unfortunately, most of it is in Swedish.

[Addition by Djerassi on 18 September 2011]: In that case, have them translated into English rather than expecting me to write biographies of your favorite scientists. I have never written any biography, perhaps because I learned how difficult that is from a master of that genre, my late wife Diane Middlebrook. Other than autobiography, I have chosen to focus on fiction and plays, but of the special subgenre *science-in-fiction* and *science-in-theatre*. Surely, every author is entitled to choose his/her metier.

My latest book that appeared in October 2008 is entitled *Meteors*. It contains three longish essays on Bela Bartók, Seymour Benzer and Barbara McClintock. I wish I could send them to you, but the language barrier is impenetrable.

Why is the book called Meteors? Because Zoltán Kodály, in a remark made during Stalinistic Rákosi regime in Hungary after WW II, contradicted the contention of the communists. Their propaganda said that Bartók, had he not died in September 1945 in the US, would have come home and would have played an important role in politics. Kodály said that this would have not happened, because "you don't use a meteor to illuminate the streets."

I chose these three persons on the basis of my admiration and love for them. Writing the epilogue, I discovered, however, that I have chosen three very similar personalities. The workings of the subconscious...

The "tribal characteristics" that you write about do exist in many scientists – although I would not call them vast majority – but they also do exist in the vast majority of people in general. This is not what makes science or scientists different or unique, just as the gastrointestinal tract or the reproductive organs do not make them unique.

[Djerassi's comment, written on 18 September 2011 after rereading that letter from Klein:]

What makes the vast majority of scientists (though perhaps not you) different is expressed in the following sentence taken from the last paragraph of the Afterword of my novel *Cantor's Dilemma*: 'Publications, priorities, the order of the authors, the choice of the journal, the collegiality and the brutal competitition, academic tenure, grantsmanship, the Nobel Prize, Schadenfreude—these are the soul and baggage of contemporary science.'

The fact that this novel has been translated into nine languages and is now in its 27th print run; and that I continue to get letters from young scientists telling me that they wished they had read that book earlier in their career must tell you something. But perhaps you should read a recent book by a fellow Hungarian, István Hargittai's *Drive and Curiosity*: What fuels the passion for science? (Prometheus Books, 2011). In it, he collected over a dozen interviews, amplified by literature research, with actual or quasi-(i.e. non-anointed) Nobel laureates which he has woven into an intriguing account of how great scientific discoveries are made. I would say that the majority of them display also the characteristics I describe in my novels and plays.

My point is that if we want to tell non-scientists how science and scientists work, we should tell them about what is different. How can we otherwise inspire young people to choose science for the future? Romantic or not, *Arrowsmith* has avowedly influenced many youngsters to choose science and some of them became great scientists.

It is not romanticism, idealism or illusion that turn these people on, but the excitement of discovery, the exploration of the unknown, the wish to understand this amazing world that is outside and within us and is not limited to our petty social interactions.

Is it the sum total of many trivial patterns that influence the impact of science? In many fields, it does. It is probably true for many of the most important inventions and applications (it is not true for PCR, because that was actually invented by a madman, Kari Mullis, just because his new girlfriend fell asleep and he had nothing to do while waiting for her to wake up. But Kari is not an example for anything except himself.)

I understand that you wish to wash dirty lab coats in public, but I don't want to focus on the clean lab coats. I want to focus on the people who don't put on a lab coat at all.

How many working scientists "manage not to dirty their lab coats"? But, again, whether they do or do not succumb to that is as uninteresting that they have to eat, drink, go to toilet and have sex. *So does everybody else.*

7 February 2009
Carl Djerassi to George Klein

Dear George,

I am truly looking forward to our "Cross Talk" as Gruenzweig will call it which is fine with me.

You said something in your last letter with respect to *Arrowsmith* which reminds me how Josh Lederberg and I recalled the effect that Paul de Cruif's *Microbe Hunters* had on us. *Arrowsmith* is a much better written book, and both fulfill the purpose you mention. But this is also the point where I depart from your view: I did not choose to write a book with the purpose of stimulating young people to go into science. If that had been my motivation, I would have written something else.

Instead, there were several other motivations, one of them a form of continuing autopsychoanalysis. But let's have that crosstalk in person! And of course the other topics you suggest are also fine with me.

25 April 2009
George Klein to Carl Djerassi

I enjoyed seeing you. You are as sharp and committed as always. I also enjoyed our conversation, although I am concerned that it may have left you with mixed feelings. For me, the great "moment of truth" came when you have immediately diagnosed the phenomenon of flow which you may or may not have heard about before (?), working in yourself when you are writing your novels and plays. That, actually, proves my point. Seeking flow is what drives most humans, most of the time, whether they know it or not.

There is another point about flow that I did not have time to make. The euphoria we experience in flow leaves us with the wish to repeat it. But on repetition, the autotelic reward (Csíkszentmihályi's term for the reward from within, the euphoria, in contrast to the external reward) is much less and the third time it may have disappeared altogether. In order to reach euphoria, we tend to increase the challenge or to look for new challenges. That makes us, humans, a creative, innovative species.

Csíkszentmihályi has written a book about people who can maintain the flow in advanced age. You would fit perfectly among his objects of study.

One more reply that I did not have time to make: One of the speakers discussed why the Pill did not get a Nobel Prize. As I always try to explain (with very variable success, I must admit) the Nobel Prize is not a judgement by the High Court of Science, but a typical committee job. For that reason, the simple "why" is meaningless, it is much more the other way around. How could someone who got it actually get it, what forces moved the very different members of the committee? It is a little bit like a presidential election, except that the actors are absent.

I hope you really meant it – in spite of what you must have felt as a disappointment – when you said that we should carry on a long conversation on a Greyhound bus. I would certainly be all for it. We have already lived long lives, and neither of us knows what is left (even though we both seem to be adamant to pursue the flow until the last moment). So, the sooner the better!

5 May 2009
Carl Djerassi to George Klein

I was very moved that you took the trouble in your busy professional life to travel to Dortmund and to engage in a stimulating discussion related to my writings. I am sorry that I could not write sooner but I was without computer in Dortmund and then on lectures in Bremen and Berlin and really am just surfacing from the avalanche of messages that awaited me.

I don't think that we can extend our debate fruitfully by e-mail, but it would be great to take with you the hypothetical bus trip (*without* e-mail connections) from San Francisco to New York or perhaps just from Stockholm to Kiruna and back. But I must answer briefly (in red) between your lines in your e-mail to Walter about one fundamental point, namely that you are quoting me selectively. You have read only a smallish portion of what I have written and picked on a point which you discuss in a pretty black and white way, using partly Csikszentmihályi's ideas which are often stated in a form of psychobabble. I participated a couple of years ago with him in a 2-day long round table in Austria and while some of the audience was entranced as if by a guru, Diane (who was then still living and came with me) and I were rather critical. Not of the fundamental idea, but the preposterously guru-like manner in which this idea of "flow" is couched as if some fundamental truth had been discovered. The topic of rewards vs. intrinsic satisfaction in science is much better covered in the many articles and books by the late Robert Merton.

I have some specific comments below that are relevant if you wish to follow

Walter's (and also my enthusiastic) wish to write something for a possible published version of the symposium.

5 May 2009
George Klein to Walter Grünzweig

(on which Carl Djerassi commented the underlined interspersed between George Klein's sentences on the same day he received the copy)

Thank you for your letter of May 1. I am glad to hear that you liked our dialogue. I have not yet heard from Carl and I can only hope that he was not taken aback too much by the critical part of my remarks. Actually, as I think of it, we have touched something important and fundamental that we (or at least I) did not have the time and the reflective depth to develop, on the spur of the moment. It goes like this:

Mihaly Csikszentmihályi distinguishes between autotelic reward, essentially the euphoria generated by an achievement in flow, and external reward like prizes, promotions and all other outward parameters of success.

People tuned to flow do what they do for the autotelic reward. The external reward that may or may not come, is not in the center of their attention. They want to repeat the flow experience and, in order to continue to experience flow, they increase the challenge, improving and learning all the time.

Csíkszentmihályi has often been asked by industrialists <u>In terms of motivation by rewards, a very different group from academic researchers for whom financial rewards play a relatively small role.</u> to talk to them about flow. Some of them got so excited by his concept that they introduced prizes and other rewards for employees who would design flow inducing exercises for the workers. By doing so, however, a totally different kind of people were attracted: those who were seeking the external reward, not the flow. This transformed the psychology and sociology of their enterprise in a less than favourable direction. The real flow seekers shy away, or turn their back on it.

This is, mutatis mutandis, what happened to modern science. It is not a good development. <u>I would be interested what "modern" means to you in this regard (I mean time-wise), since I have written quite a bit about scientists 200-300 years ago (in the plays OXYGEN and CALCULUS) when single authorship was the norm and what motivation meant then.</u>

Djerassi's books and plays although well written and often very entertaining, miss this aspect.

As I already wrote you earlier: this is unfair since you have only read a small portion of my books and plays. Obviously, I don't expect you to have read them all, but in the absence of doing so, you cannot make such a generalization. For instance in my novel *NO*, the two main characters, Renu Krishnan and Celestine Price demonstrate very different motivations which in psychobabble could be described as autotelic and external. Even in my scientific autobiography, *Steroids Made it Possible* (American Chemical Society, Washington 1990) you find virtually only "autotelic" motivation. I also do so in the cortisone chapter (No. 4 entitled "No depression") of my general autobiography (*"The Pill, Pygmy Chimps, and Degas' Horse"*). Finally, a very important point is that the motivation that is less often discussed and which can also have pernicious as well as wonderful consequences is "peer approval" in science, where by the nature of the scientific tribal culture it plays a much greater role than in other disciplines. In my play *EGO*, I have the main character call it "productive insecurity."

But enough of my defensive words. The fundamental difference between us, which I totally accept, is that I have chosen to emphasize the grayer aspects of scientific behavior than the rosier ones you prefer. The purpose of my themes is not to write books to encourage young people to go into science – many people have done so and have done it very well – but to illustrate some of the grayer and less attractive aspects. This does not mean *at all* that I recommend such behavior, but it is absurd to try to hide it.

They see scientists as focused on the external reward, exclusively. This is my criticism. Even though much effort in science may be, by now, as he describes it, this is not what we should strive for. On the contrary, we need to save that endangered species who do the science for its own (and their own flow's) sake.

Concerning your question, I fully agree with you that a discussion of this type cannot be reproduced or revived in retrospect. This does not mean that I am totally negative to writing something, but, then, more in a way of commentary, focusing on what has been said above.

6 May 2009
George Klein to Walter Grünzweig

I am somewhat familiar with deconstruction, Derrida, Foucault and others and have written several very critical texts about attempts to "deconstruct" scientific texts. I was particularly harsh in my criticism of a book by Fujimura, published by Harvard University Press, entitled "Crafting science." Unfortunately, my critical texts exist only in Swedish.

My review of the book starts with a metaphor. Imagine a stone deaf sociologist who has never heard a sound, but who undertakes the task of examining the sociology of the opera. He assembles a large number of facts, most of them quite correct. Since he does not have the faintest idea about the ways in which music and particularly the human voice can influence our feelings and thoughts, he draws the conclusion that the success of an opera, the reputation of an opera house and of the singers is exclusively determined by networks of power, money and influence.

The book claims that scientists are obsessed by an "illusion of reality." She rejects this totally. Watson and Crick has not discovered the structure of DNA, they have "formulated" the structure. There are no discoveries, only claims, no text has any connection to reality, only to other texts.

I would like to see the deconstructionist whose child gets pneumonia but who, instead of giving him or her penicillin, starts deconstructing a text about the action of penicillin. Until I see that, I remain a sceptic, as far as the deconstruction of scientific texts is concerned.

Having said this, I can of course see the value of well reasoned literary deconstruction. Did it not start out as a useful concept, particularly by Derrida, but was later hollowed and politically distorted?

I can see what you mean when you speak about Carl's construction of the scientists' world as a culture. Yes, indeed! But this is exactly what I am critical about. The success mindedness inundates more and more, but this is not what science is all about. Yes, it is part of it, but it is the other part that is worth living for.

If you have time, I can warmly recommend the short, popular version of Mihály Csíkszentmihályi's book *Flow*. It argues the case much better than what I can do.

7 May 2009
George Klein to Carl Djerassi

Let me first comment on what you write about Csikszentmihályi and flow. This is now another case where style and social interaction are allowed to interfere with content.

Csikszentmihályi is not a good lecturer. He does not come out too well on discussions either. It is also true that he has many epigons (including myself, an amateurish one) and once you have penetrated what he is talking about, you are tempted to regard him as a guru. But this is not important, because:

What Csikszentmihályi says is not "psychobabble." On the contrary, it is experimental, observational, and statistical psychology, based on research of unusual profundity and a vast amount of data. Most of it is published in specialized journals with their professional jargon and we who sit as his feet tend to oversimplify what, from his point of view, are already oversimplifications.

In my experience, the flow concept generates an immediate "aha" reaction in most people when you tell them about it. A curious exception is the real, absolute flow-addicts who may respond negatively, like a fish who takes it so self-evident that he lives in water that he regards every mention of it as trivial nonsense.

I have written a whole book about flow addicts. It is one of my few books that has been translated to English. I would like to send it to you without any expectation on my part that you will read it. Where can I send it?

I know at least some of your writings about scientists of bygone times. *Oxygen* is a good example, with the Nobel Committee that gives a prize for past achievements. Incidentally, having sat in a Nobel Committee for three decades, I can tell you that the committee discussions as in *Cantor's Dilemma* or in *Oxygen* don't have the faintest resemblance to what the discussions in the Nobel Committee are like. But that is beside the point. The main point we are arguing, and the only one worth arguing about, is still this:

Yes, scientists are human. Yes, they can show all signs and symptoms of human frailty. But that is not what makes them interesting, just like Beethoven's of Schopenhauer's acrimonious quarrels with their landladies is not what you want to hear about in the first place.

What you (or at least I) want to hear about is the exhilaration and, indeed, the euphoria of the human mind when it uses its enormous creative potential (of which only a small fraction is being used most of the time) in understanding some of the world within or outside us, or, indeed, meeting the major existential problems we all have to meet in poetry, music and art. The social interactions and role playing of scientists is nothing but a boring monkey game.

Another point: I totally agree with you that peer approval is very important. One can also talk about "group flow" in certain contexts. But even in this context, I believe that the peer approval you (or at least I) most appreciate is coming from those who appreciate your genuine commitment, not just your role playing.

I shall not give up the hope of a long conversation. But if not Stockholm-Kiruna, why not Budapest-Vienna? Is a ship on the Danube not a better milieu than a bus?

Weak Bonds – Strong Effects:
How Carl Djerassi Left Chemistry and Became a Writer

Henning Hopf

Chemistry has its problems with being appreciated by non-chemists. Chemists often feel disappointed, even upset, that the general audience does not understand them; lay-people often regard this particular natural science with a mixture of non-understanding, bewilderment and distrust, even outright opposition. Probably no other person has done more to bridge this gap than Carl Djerassi. In my contribution to this volume I want to look at Carl Djerassi from the standpoint of a practicing chemist. Although I am officially retired, I am still working, publishing, lecturing in an area Carl has left many years ago: academic organic chemistry.

Chemistry is the science of the metamorphosis of matter. As such it analyses matter, as we find it around us both as 'living' matter – in plants, in animals, including ourselves – and as 'dead' matter everywhere in the universe – in minerals or rocks on Earth or on other planets. All of these chemical compounds may be called 'natural products.' Analysis is the major task of many other scientific fields, but Chemistry is more: it is also the science of transforming matter. As August Kekulé, one of the forefathers of Organic Chemistry, pointed out, its subject is not only the given (i.e. natural) compound, but also what can come out of it, how it can be transformed. Therefore, chemistry is a creative science of the first order; its practitioners need – in Robert Musil's words – a *Möglichkeitssinn*, a potential meaning as much as a *Wirklichkeitssinn,* a meaning pertaining to reality. (Musil 16)

To be the subject of a chemical study, the building blocks of matter – the chemical elements – have to form *bonds* between themselves that last at least as long as the time required to look at them with a physical or chemical method. In fact, the method must be faster than the compound or event it looks at. Chemistry is hence the science of bond-making and bond-breaking, of forming new bonds and dissolving old ones.

This has been noted many times before, also by people who are not from the field of chemistry such as Johann Wolfgang von Goethe in his famous novel

Wahlverwandtschaften (Elective affinities). In German literature, this type of book is called a *Bildungsroman* – but I call it a *Bindungsroman*, a bonding-novel. This is exemplified best by its most famous quote – at least in the area of chemistry – by the so-called *Chemische Gleichnisrede* which forms the basis of this novel about a scientific experiment with human beings:

'Wenn Sie glauben, daß es nicht pedantisch aussieht,' versetzte der Hauptmann, 'so kann ich wohl in der Zeichensprache mich kürzlich zusammenfassen. Denken Sie sich ein A, das mit einem B innig verbunden ist, durch viele Mittel und durch manche Gewalt nicht von ihm zu trennen; denken Sie sich ein C, das sich eben so zu einem D verhält; bringen Sie nun die beiden Paare in Berührung: A wird sich zu D, C zu B werfen, ohne daß man sagen kann, wer das andere zuerst verlassen, wer sich mit dem andern zuerst wieder verbunden habe. (Goethe 276).

'If don't think it's overly pedantic,' the Captain replied, ' I can briefly summarize what I mean in symbolic language. Imagine an A which is intimately connected with a B, and which many means or several powers cannot separate; imagine a C, which has an identical relationship to a D; now get these two couples in touch with each other: A is going to throw itself at D, C at B and one will not be able to say who left the other first and who was first to take up again with the other.

Expressed as a chemical equation, and excluding all the poetry of Goethe's language, this reads: AB + CD = AD + BC.

This equation describes among other events: four acts of adultery leading eventually to three deaths. A chemist would call this a process with a high yield. In chemistry, this transformation is called a "double transformation" or "metathesis" ("Doppelte Umsetzung" in German). As in any chemical process, bonds are destroyed and bonds are formed. Incidentally in Goethe's novel – written in a wonderful, non-pedantic language – the person who wants to mediate between the protagonists has the name "Mittler" (mediator). In chemistry, we call this a catalyst – an agent making bond changes easier.

What are these bonds – in chemistry, not the ones between human beings? As so often with rather complicated matters, we make our life easier by categorization. A chemist distinguishes between strong and weak bonds. With the strong bonds, the skeleton of chemical compounds is formed, they are the bones. The weak ones are often responsible for the *function* of chemical compounds. They are the flesh, the muscles, possibly even the psychological forces. The strong bonds are the hardware, the weak bonds the software. A compilation of various types of chemical bonds is shown in the diagram below for which I am endebted to Burghard König at Universität Regensburg. Without going into details, it becomes obvious that there are very strong bonds (high numbers), which are difficult to break, and there are weak ones (small numbers), which are actually so weak that a molecule hardly 'feels' the presence of another reagent, of another potential reaction partner. It is like two persons passing and merely glancing at each other. To destroy strong bonds requires high energies, i.e. high temperatures. This happens as we burn a substance or during cooking, especially frying. To destroy a weak bond, much weaker forces are needed: we can pull out a hair with hardly any force. Without changes in the character of the bonds there would just be the approximately 100 elements we find in the universe – certainly not what is most important for us: life, especially our own personal life.

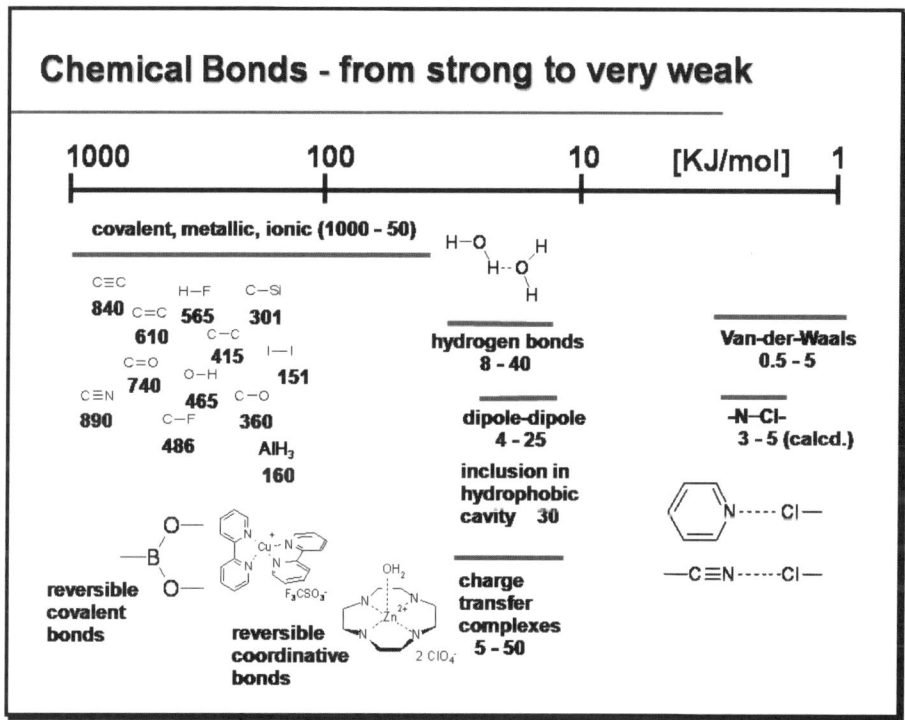

The look at chemistry as described above is far too simple, too static. Bond changing processes are actually dynamic processes, they occur in time. Can we find descriptions of the dynamic behavior of the bond switching process in the chemical literature? Yes, indeed, and a multitude of Nobel prizes have been awarded to scientists who have investigated the time dependent nature of chemical processes.

One example of such a process is a chemical equation that describes what we call a "solvolysis reaction" in which a chemical compound reacts with the molecules of a solvent or reaction partners added to that solvent.

We have a starting material which has a least one critical bond, the R-X bond, a bond that 'wants' to react in a certain manner. By stretching it, it breaks and

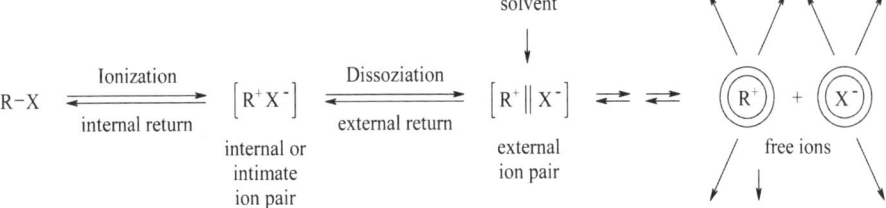

a so-called "internal" or "intimate ion pair" is formed. Obviously, the process, "ionisation," can be reversed and by the so-called "internal return" the starting situation can be regenerated. However, the original partners have another possibility: they can dissociate, allowing solvent molecules to slip in between them. This process is reversible as well – in the "external return" the separating partners remember their original half. Very often, however, the inevitable happens: both former halves surround themselves with their own wall of solvent molecules and become completely independent: they are free ions, and a return to the starting situation has become extremely unlikely. Now they are free to react with any other reagent around: they have nothing to lose any longer. As far as the bonding forces are concerned – in the starting situation we have strong bonds but they become weaker and weaker as we proceed from left to right.

The relationship of the above solvolysis scheme to the *Wahlverwandtschaften* is obvious – as it is obvious that the language of natural scientists is much, much less powerful than an author's such as Goethe. Roald Hoffmann and I have observed in a recent essay that scientists often think that words do not matter but that equations, formulas, and spectra do. But facts are mute; without words, no sense could be made of this world. Words humanize the inanimate world, form a connection, a bond *with*, a bond *to* a human being. Words mislead much less than they encourage, for it is just through their anthropomorphism that they provide a rationale for the often tedious work of the chemist. In other words, if one understands the science that forms the basis of a phenomenon, then colloquial, anthropomorphic, colorful expression makes inanimate matter spring to life. Still, scientific language has its limitations.

In fact, novelist Thomas Bernhard, a fellow Austrian of Carl Djerassi's, expresses this very well in his novel *Verstörung* (1967, published in English under the title of *Gargoyles*): "Substances, something immensely Chemical. The greater the distance to the conventional notion of nature, the more beautiful, the more powerful, I want to say: poetic." ("Substanzen, ein gewaltiges Chemisches. Je weiter es sich von dem konventionellen Naturbegriff entfernt, desto schöner, gewaltiger, ich will sagen, poetischer." Bernhard 98) The question raised here is a very interesting one. The language of organic chemistry, the "nomenclature" as chemists say, follows strict rules, written down in the so-called Geneva Nomenclature Rules. With these rules, one can 'construct' a chemical term in a very rational way. Modern chemistry unfolds much faster than these rules can be developed and adopted. The consequence is that more and more chemical names, especially for chemical compounds, are given in what is known as "trivial nomenclature." Trivial nomenclature uses comparisons, metaphors, and images from daily life; it is a strongly anthropomorphic language. And it may well be that one of these days – actually rather soon – chemists will communicate with each other not by talking but by pointing to interesting and important graphical details which they see on a screen they are looking at together – a conversation consisting of a very simple spoken language and interrupted by many "Ah's" and "Oh's" and "I see's." It will be like standing in front of a painting and talking about it.

What does this all have to do with Carl Djerassi? My thesis is that Djerassi has been moving along this "bonding scale" from strong to weak in a very consistent manner in the course of his life. There is no abrupt split between his work as a chemist and his work as a writer, poet, and essayist. There is no *before* and *after* – but a continuous development.

When Carl Djerassi was a practicing chemist, he originally worked in organic synthesis, a field which he now sometimes calls "macho chemistry." (Carl Djerassi, private communication) This image is not inadequate, but it only tells part of the story. Synthetic chemistry is the field of the strong bonds as we defined them above. The compounds he was then interested in were the steroids. And in all the ensuing work on other classes of natural products – alkaloids, antibiotics, lipids, and terpenoids – the bond making and bond breaking, mostly between carbon and carbon but also between carbon and so-called hetero atoms – oxygen and nitrogen atoms, for example – constituted the main part of his work. Compound making is hard work; it is like finding a pathway through a thick primeval forest and besides endurance, the most important qualities one needs are fantasy and courage. These characteristics are required because there are so many failures. What do you do if you reach a seemingly insurmountable peak, a wild and mighty river you have to cross, or a deep abyss? Find detours, find solutions for difficult situations, develop completely new approaches. And try, time and again.

Natural product chemistry is carried out for various reasons. The most important one is to understand certain biological processes, especially the really important ones, sensual perception, understanding our senses, reproduction. It is obvious that this work cannot be done by chemists alone, but surely they play a most important role in this promethean game that has now reached the molecular level. It is also done for a plethora of medical reasons, for finding compounds in nature that can fight diseases, for the manipulation of important biological processes.

Some of the steroids I just mentioned are of crucial importance in controlling human reproduction.

In this area it is not only the strong bonds that count, but also, and often to a decisive degree, the weak bonds. Many natural products have a biological function and to exhibit that function they have to engage in weak bonds – hormones, enzymes, all kinds of receptor and container molecules. Just stable enough to cause an effect, but not so stable so that this effect remains 'switched on' forever. Today we call the chemistry which investigates the making and breaking of strong bonds "covalent chemistry." And we speak of "supramolecular chemistry" when we are dealing with weak bonds, the bond types in the right half of the diagram above.

When Djerassi worked on the organic compound that eventually should lead to what we colloquially call "the Pill," the term "supramolecular chemistry" had not been introduced yet into the chemical discourse. There were several reasons responsible for this, the most important one being that the methods used in traditional chemistry were not subtle enough. To investigate subtle, yet extremely effective bonding processes, processes which have a biological consequence, the corresponding subtle technology is required. In other words, you cannot repair a tiny and intricate wrist watch with a sledge hammer. The introduction of these 'softer' methods – they all came from physics – into chemistry began around 1960 and Carl Djerassi was involved in this process from the very beginning. Whether he was actively searching for 'softer' methods at that time, I do not know. But he recognized their future importance very early. In fact, at first he started with a relatively 'brutal' method: mass spectrometry. In this analytical method originating from physics, a molecule is energized to such an extent that it fragments into numerous bits and pieces. These pieces, their mass to charge ratios to be exact, are registered, and from them the structure of the original molecule is deduced. It is like putting together an antique vase or statue from countless shards. Mass spectrometry in its original form is a so-called destructive method. The original material one wants to investigate is not available at the end any more. The object of nature has been translated into another form of representation, a formula, a structure and, finally, into information and eventually new knowledge. In the last few decades, mass spectrometry has also become softer, less brutal.

With the application of physical methods to organic chemistry, Carl Djerassi began to show an interest in the other half of chemistry, its other face I referred to earlier: analysis. Although he was still what we call a "synthetic chemist" at that time, his bonds to this field began to weaken. And again we see this change from hard to soft, from destructive to non-destructive methods, from invasive to non-invasive, in a sense from 'male' to 'female.' The 'softer' methods he used after his pioneering work on mass spectrometry, which had a strong influence on other chemists, natural scientists scientists and even instrument makers, have the technical names of Optical Rotatory Dispersion (ORD) and Magnetic Circular Dichroism (CD). It is not important here to understand the physical principles behind these methods. What is important is the means used to disturb the molecules under investigation – how the energy applied is of electromagnetic nature, i.e. light is used. The molecules about which we want to learn something – most often the way in which its atoms are bonded to each other, its connectivity and spatial structure – are tickled by light. Depending on the very nature of this interaction the excess energy given off once the exciting energy has been turned off is an electromagnetic signal which can be read by an appropriate detector device. Again, physical properties have been translated into another type of language and scripture, but this time the object studied survives the process physically unharmed, the method is non-evasive and non-destructive. It is much softer than the methods discussed so far, which can be very important. At the beginning of a study, many medicinally active natural compounds are only available in minute amounts. It would be a serious loss to destroy them, even if this loss is a "creative destruction" in Schumpeter's sense.

Assuming that the development of a scientific career follows a logical pattern – what would be the next step of such a development, a development which we could call a "dematerialization of chemistry"? Is it possible that the science of matter par excellence is beginning to lose its material basis? That this down-to-earth-science is becoming a world of signs, icons, and symbols? That reality as we find it around us is replaced by virtuality? Yes, this is conceivable, and it is interesting to see how Djerassi approached this problem and solved it for himself.

Chemistry is indeed more or less another language to describe and make sense of the world surrounding us (and in us). There are many more languages: those of the other natural sciences, medicine, psychology, law, sociology and the various arts. All of them are different 'windows' allowing us to view the world from different angles, from different standpoints. The notion that there are only 'Two Cultures' – of the Natural Sciences and the Humanities – is an all too simple, actually often misleading, categorization!

When you want to terminate your relationship, your bonds to chemistry – for whatever reasons, under whatever influence and circumstances – there are two roads that are open for you. Leave the material world of laboratory chemistry and go into virtual chemistry, or try to quench your thirst for language, for talking, for interaction, by leaving the puzzles of the material world completely and concentrating your efforts on human beings.

It is interesting that Carl Djerassi tried the first route first. Not many people, not even among chemists, know that Carl Djerassi made a brief foray into computer chemistry when the first digital computers became available, from the invisibly small to the materially non-existent. Chemical compounds represented by ones and zeros in a computer – whether we are talking of their toxicity or their smell, their color or weight. Many of the older methods of structural elucidation involve algorithms, they are more or less intricate algorithms. That is why they are intellectually so enjoyable – it is like solving a complex criminal case where you have to put the stones of a mosaic together to have a complete picture in the end. But the artificial intelligence and its uses in chemistry that Djerassi became interested in for a short period of time are more difficult than solving the structures of chemical compounds by physical methods.

It would be interesting to speculate how this new field would have developed in his hands. My guess is that it would have flourished, particularly because the situation was so very similar to Djerassi's previous work: he had recognized the importance of a new, rapidly evolving new area of chemical research, he belonged to the avant-garde of that field and – also very important – from his previous work, which had resulted in over twelve hundred original articles and seven monographs dealing with the synthesis and analysis of organic chemical products, he knew how overwhelming the difficulties can be when a new research field has to be established.

As we all know, Carl Djerassi decided not to investigate the nature of chemical bonds anymore, whether these were strong or weak. The bonds, for whose discovery and description the forces of the heart, of feeling, of psychology and foremost a more colorful language than the language of chemistry is required, became increasingly important to him.

At the very root of Carl Djerassi's development and unbroken creativity is his interest in bonding, in understanding bonding forces. To be successful in chemistry, in any field, you have to develop a very strong affection to this field. How this affection develops is an unsolved problem; it is obviously a process influenced by many parameters. If you cannot establish this bond, it is extremely unlikely that you will be able to endure the frustrations, set-backs and failures of a scientific career. Although this solid basis is indispensable, in my opinion, it can be – and often is – very destructive for further developments. This is why many scientists stick to whatever area they have discovered for themselves and become experts in a certain field. Scientists often praise themselves for their rigor – but sometimes this rigor is not very different from *rigor mortis*, into which it passes over easily at the end of a scientific career.

Carl Djerassi was and continues to be very different. Clearly, he has been able to establish these most critical bonds, otherwise he could not have celebrated this distinctinve scientific success. He was able to develop vastly different research areas, a spectrum of interests only achieved by very few scientists, especially if one considers the enormous political and practical impact of his work. However, at the same time he was also capable of breaking these bonds, of letting loose.

Had he continued in the steroid field he would – no doubt – have been one of the leading natural products chemists of the world. He had the courage, the flexibility, the vision to break his bonds to this particular field. This process, about which we also do not know very much, is at the root of creativity, I believe; not only in chemistry. When you leave a field, you also leave behind you the people who are active in this field. In several of his novels, Carl Djerassi has talked about the tribal secrets of chemistry or those people who practice it. Yes, this is very true. We are a tribe, or tribes, rather, and often tribes dominated by male chauvinism, machismo. The whole 'in group–out group' scenario which is so prevalent in the extremely competitive science of chemistry, proves this.

Obviously, for Carl Djerassi, this was a world in which he did not want to spend the rest of his his life. He did not leave science in the sense of quitting, but bonds to other areas were stronger. And how and whether he contributed to the eternal human quest to understand the world on a molecular or – let us say – human-sized level finally does not make such a big difference. The willingness to participate in the project we call Enlightenment is what counts most.

Although I read many of Carl Djerassi's scientific papers already as a graduate student (interestingly enough at the University of Wisconsin in Madison where he had received his Ph.D. about 20 years earlier), I only met him personally in the early 1980s when he already had left the 'hard' sciences. This was at a reading from his novel Cantor's Dilemma *in a bookstore in Braunschweig, during one of his many reading lectures that led him to numerous towns in Germany. From then on an intense relationship developed and we met many times on both sides of the Atlantic, especially since my first stay as a visiting professor at the Chemistry Department of Stanford University in 1997. The recurrent theme of our long conversations and of our correspondence has been the relationship between the sciences (i.e. chemistry) and the arts, i.e. fiction, poetry, and language. We both agree that creativity forms the basis of science and the arts, and that, hence, a Two-Culture schism does not exist. – See also my recent biographical note on Carl Djerassi in* Nachrichten aus der Chemie, *2009, 57, 22.*

Works Cited

Bernhard, Thomas. 1985. Verstörung. Frankfurt/M.: Suhrkamp.
Goethe, Johann Wolfgang von. 1981. *Die Wahlverwandtschaften*. Hamburger Ausgabe. Vol. 6. München: Beck.
Musil, Robert. 1978. *Der Mann ohne Eigenschaften*. Reinbeck: Rowohlt.

Carl Djerassi:
Scientist – Artist – SciArtist

Christian R. Noe

Carl Djerassi was the second person to be elected by the Austrian Academy of Sciences to receive the newly established Ehrenring (Ring of Honour) of the Academy in 2008. The ceremony took place in the Festival Hall of the Austrian Academy of Sciences located in the building that formerly housed the University of Vienna. The ceiling of the hall is decorated by a magnificent fresco with the Allegory of the University. Although kept in traditional ecclesiastical style, the painting reflects the spirit of Enlightenment of the university reform in the mid-18th century driven by Franz Stephan of Habsburg-Lorraine, the husband of Empress Maria Theresia. The characteristic features of the four Faculties are shown in a very subtle way: "notitia (rerum divinarum)" for theology; "(causarum) investigatio" for philosophy (at that time including natural sciences); "(iusti ac iniusti) scientia" for law; and "ars (tuendae et reparandae valetudinis)" for medicine. "Ars," very much in contrast to modern(ist) conceptions, was at that time by no means an end in itself. The four steps of "taking notice" of a problem, "searching" for an answer, "knowing" the facts and "acting skilfully," describe the full scope of the scientific process. This definition is far beyond the present reductionist perception of science limited to research ("investigatio") only. This festival hall was certainly the perfect place to honor a personality like Carl Djerassi.

Carl Djerassi has been active and visible in the public for more than half of a century. He is one of the most prominent natural scientists globally, known also to large number of people not specifically interested in scientific questions. His prominence in science in connection with his interest in art provokes a comparison with Pablo Picasso, the prominent artist who had also been 'visible' in the public for a long period of time. In the general public, both of them are well known for only one segment of their achievements. In the case of Picasso, this is his rather abstract way of depicting doves and women. Only experts in art know that he has been influential in most of the art movements of the first half of the 20th century and that he was also a very 'political' personality. In the case of Carl Djerassi it is his role in the invention of the "Pill," which has made him popular. However, only experts in organic chemistry know about his other scientific achievements and a very different community is aware of his contributions to literature. The two of them have in common that they have influenced society much more than generally perceived. Obviously, celebrity has a fundamental draw-back, because the personality is reduced in the public perception to one specific achievement to allow easy recognition by direct association of the person to one aspect. This limited view obscures a series of important other achievements of these two exceptional men.

Carl Djerassi is known to the public as the "inventor of the pill," or, in his own, preferred, definition, as the "mother of the pill." (see Djerassi 1992) Driven from his native Austria in 1938 by the Nazis, Djerassi arrived in the United States via Bulgaria to finish his studies in record time. In very young years he made an important and remarkable decision that eventually made him world famous, namely to join the rather small company Syntex in Mexico City. This specialized company had been established to synthesize steroid hormones from extracts of Mexican plants. At that time steroid research was what we nowadays define as cutting edge research. On the one hand the best chemists were challenged to expand beyond the frontiers of their knowledge and their synthetic and analytic capabilities. On the other hand the corticoids, a class of hormones, had just attained specific therapeutic interest and opened new doors to treat inflammation and to modulate immune response. Progesterone, the pregnancy protecting hormone, was the central product, both in the biosynthetic pathway, which starts from cholesterol, and in industrial manufacture of corticosteroids. Progesterone and hormones derived from it were in the focus of the company which Carl Djerassi joined in 1949.

Already long before, in 1924, Ludwig Haberlandt, a Professor of Physiology at the University of Innsbruck, had found that progesterone prevents pregnant women to conceive a further child during pregnancy. (see Haberlandt 55-67) Based on this observation, work on influencing pregnancy by progesterone had been taken up already before the Second World War. At that time, Schering in Germany and Gideon Richter in Hungary were companies active in this field. Thus it was not really surprising that Syntex, a company dealing with progesterone and its derivatives, embarked after the Second World War on a completely new field of drug application. Oral contraception was promoted by activists like Margret Sanger or Katharine Dexter McCormick to allow women more freedom in the decision for pregnancy and to reduce their dependence on men. The coming role of the oral contraceptives as life style drugs which would change society could not be envisaged at that time.

Carl Djerassi's self-definition as "mother of the pill" indicates that for a scientist, in particular for a synthetic organic chemist, a research project is not just a technical matter but has a much to do with the timeless Promethean dream of 'creating' something new. For a synthetic scientist, the moment of having a desired molecule in hand for the first time resembles emotionally a little bit seeing one's newborn baby. It was certainly tempting for Carl Djerassi to connect motherhood with the aspect of giving birth to something that prevents pregnancy and allows planning of pregnancy. By designing the active drug ingredient norethisterone, he gave birth to the first compound allowing oral contraception.

There is something very peculiar about organic synthesis. Born from pre-scientific alchemy, chemistry evolved in the 19th century based on the concept that matter is made up of atoms. It soon became apparent that organic molecules are made up of combinations of specific atoms, above all carbon, oxygen, nitrogen, phosphorous and hydrogen. Nevertheless, it took several generations of chemists to understand how these atoms are linked to each other. Towards the end of that century, something almost miraculous had emerged, namely the modern formula system of chemistry providing a language for the constitution of the 'molecules of life.' The principle is rather simple, comparable to a kind of Lego game in which each element undergoes a defined number of interactions in a defined spatial arrangement. Total synthesis of such a molecule became the established way to prove its constitution. It is not surprising that the increasing precision of the formula language also laid the foundations for a rapidly growing chemical industry. Organic chemistry became a leading science.

Nowadays we know exactly that the chemical formula language is not just a useful tool to understand the composition of molecules and to carry out their synthesis. Chemical formulae are much more. They are in fact very precise pictographic representations of the shape of very tiny entities, the molecules. The more the reliability and power of the formula language was proven, the more organic synthesis shifted from a method dedicated to the understanding of structure and reactivity of organic compounds into an art of design, mostly 'sculptural,' to achieve a specific purpose, e.g. to create a new drug or a new dye. Not surprisingly, scientists often generate new molecules in a manieristic manner, a kind of scientific 'l'art pour l'art.'

When Carl Djerassi entered the scene in the mid of the 20th century, organic synthesis was on its way to tackle rather complex molecular structures. Steroids, the structures of important classes of hormones, were in the focus of this research. But Djerassi's work did not just prove the structure of a steroid by its total synthesis. His 'design work' went a step further by aiming at improved properties of a naturally occurring hormone.

A recent poll survey among leading drug researchers on the most important achievements in drug discovery during the last 50 years included a statement that at the beginning of this phase a new era of drug research began, when scientists started to take into account biochemical and pharmacological knowledge and concepts. Djerassi and oral contraception were mentioned explicitly in this survey. (International Pharmaceutical Federation) In fact, it is amazing that the first drug compound invented for oral contraception was not an orally available derivative of the natural gestagen progesterone, which would have seemed to be the logical approach in a company specialised on progesterone. Djerassi designed his norethisterone based on the recently discovered biochemical knowledge that a testosterone derivative, exhibiting male sexual hormone activity, would change its hormonal profile into that of a gestagen, a female pregnancy protecting hormone, just by leaving out a tiny methyl group at a specific position of the molecule.

He also used the knowledge that a specific group, the ethinyl group – at another position of the hormone – would make the compound orally available. To implement this concept he used a further transformation. He took the female sexual hormone estradiol and transformed it into nortestosterone, making use of a rather new synthetic method, the Birch reduction. With respect to hormone action, this design meant a twofold 'trans-sexuality' at the molecular level. If one would compare drug design with sculpturing at the very tiny molecular scale, the design of norethisterone would correspond to a switch from the copy of the natural entity into a concept-based design of a completely new structure. The design work of the scientist corresponds very much to that of an artist: First comes the idea, then the work with the material and finally the product. This is not just a copy of the reality but reflects the concept of the artist. In the case of norethisterone, it was the perfect implementation and key element towards the advent of oral contraception.

The "Pill" reached the market in the sixties of the last century and contributed massively to the then occurring changes in society, above all the emancipation of women. It is not surprising that the huge impact of the "Pill" on society turned Djerassi into a widely known scientific celebrity. To be the "mother of the Pill" certainly would have been enough of an achievement for a whole scientific life. Djerassi, however, did not lean back but made further important scientific contributions. He became a leading organic chemist competing in his work with the most prominent researchers. With the first synthesis of the hormone cortisone from a plant, a small research group in Mexico celebrated a great triumph over Robert Woodward in Harvard, the most famous organic chemist of his time. This further increased Djerassi's reputation among his colleagues.

In the seventh decade of the last century, organic synthesis was thus a prominent science aiming at molecules with increasingly complex structures. This development was due to the introduction of new instrumental analytical methods which themselves constituted the pioneering first steps into the upcoming use of computers and robots in scientific work. It is characteristic of Djerassi that he not only used such new methods, but contributed himself to their introduction and application, mainly in structural analysis of complex organic molecules. The use of mass spectrometry, optical rotatory dispersion and circular dichroism spectroscopy in organic chemistry was significantly promoted by new rules and application methods developed by Carl Djerassi in close co-operation with the best theoretical scientists.

The project leading to the commercial introduction of the Pill was implemented with remarkable speed. Being employed in an industrial company, Carl Djerassi became quickly involved in the complex process of drug development. While many researchers at that time were only interested in 'basic science' and not in application, Djerassi acted out of the awareness that it is not enough to have a good idea or to write a paper on a finding of potential relevance, but that it is important to contribute to the full implementation of a concept. It is the step of the scientific process which is present as "ars" in the fresco of the festival hall in the Vienna Academy of Sciences. Nowadays the term "translational sciences" has been introduced for this type of activity. The importance to translate new knowledge into an invention, then into an innovation, and finally into industrial practice, is well recognized in the current world of science.

This holds especially true for the area of drug research and biomedical science given the growing gap between the soaring costs of research and many failures in the introduction of new drugs on the market. It is no wonder that at present "translational institutes" are starting to appear everywhere. No place in the present world represents this "translational" aspect side of the co-existence of academy and industry better than Silicon Valley. A brief look at its history reveals that Carl Djerassi's move to California and the transfer of Syntex thereto meant an important milestone in the formation of this unique creativity generator. (Rao) But Djerassi was not only a pioneering translational scientist. He also founded the "Syntex Institute of Molecular Biology" in the Stanford Industrial Park and thereby extended the range of his activities into molecular biology. This was certainly one of the first cases when the term "molecular biology" appeared in the title of an institution. He certainly also enjoyed the visit of King Carl XVI Gustav of Sweden to another one of his research companies, Zoecon, which dealt with insect hormones.

The visionary power of Carl Djerassi reached beyond all this. We live nowadays in the 'age of the brain' and are much concerned with artificial intelligence. In this area, too, he was a pioneer, co-operating with the best scientists in the field. He worked on applications of computer artificial intelligence techniques to organic chemistry and studied them for nearly a decade in collaboration with one of the 'fathers' of artificial intelligence, Edward Feigenbaum, and Nobel laureate and geneticist Joshua Lederberg.

Instrumental analysis, molecular biology and artificial intelligence are great themes for which Carl Djerassi did not provide the basics but in which he was certainly one of the first in successful implementation – "applied science" in the best sense of the word. The pioneering and sometimes challenging role of Carl Djerassi in translational sciences deserves more public recognition in our time when scientists are increasingly expected by society to be effective – from hypothesis to implementation – and not only to play games in their own scientific silo. Comparable to Pablo Picasso, whose fantastic vision brought emerging movements in painting to perfection, Carl Djerassi, with his enormous instinct for upcoming scientific opportunities has played for decades the role of a kind of living trend barometer for significant themes in science.

At mid-life, Carl Djerassi had achieved everything a scientist could dream of: a unique scientific record, challenging projects with excellent results, high respect among colleagues, recognition from the formal and informal public, and last not least a significant income. It is not surprising that at that time Djerassi's interest in art turned him into a serious collector, in particular of the painter Paul Klee. It is, however, rather unique that he converted a cattle ranch (named SMIP, for: Steroids Made It Possible) near San Francisco, a beautiful 1,200 acre spread overlooking the Pacific Ocean, into an artists' colony which by now has hosted over 2000 artists in the areas of literature, visual arts, music, and choreography. The "Djerassi Resident Artists Program" (DRAP) was established as a memorial to his daughter Pamela, a promising artist herself, whose life ended tragically. We may speculate that the death of his daughter induced a crisis or a self-finding period during which the artistic designer of useful molecular art began to ask himself whether a more immediately artistic work might not be more rewarding.

Diane Middlebrook, his late third wife, a professor of literature at Stanford and a well-known biographer, supported Djerassi's plan to take the full step into the world of art as a writer, first with autobiographical books, above all *The Pill, Pygmy Chimps, and Degas' Horse* (1992) but quickly developing his own field and style of writing, for which he coined the term "science in fiction." Novels such as *Cantor's Dilemma* (1989), *The Bourbaki Gambit* (1994), *Menachem's Seed* (1996) and *NO* (1998) integrate genuine scientific themes and questions into fictional texts which sometimes seem more like a biographic report than a novel.

Before long, Djerassi took the next step by evolving into a playwright. "Science-in-theatre" was the starting point. Plays like *An Immaculate Misconception* (2000) or *Oxygen* (2001, co-written with the Chemistry Nobel laureate Roald Hoffmann) still reflect the scientific environment in which Djerassi used to work. The play *Phallacy* (2005) is also connected with chemistry. The central character displays features of the author's friend, the chemist Alfred Vendl. However, in this case, as in some of his novels, the main theme deals with ethics in science. The thematic developments of his plays then move from chemistry to mathematics and to philosophy. The main characters of *Calculus* (2002) are the world famous scientists Newton and Leibniz. The overriding issue of scientific priority is in the focus of this play. At last, Leibniz is generally perceived in public more as a philosopher than as a mathematician. Djerassi's turn to philosophical themes is then elaborately taken up in *Four Jews on Parnassus – A conversation: Benjamin, Adorno, Scholem, Schönberg* (2007); an extremely ambitious work integrating intellectually demanding thoughts of philosophers into a dialogical framework. His latest play *Foreplay* (2011) also takes place in a philosophical environment, namely of Theodor Adorno, Hannah Arendt and Walter Benjamin.

It is not surprising that Carl Djerassi wants to be recognised and respected as an artist independently from his achievements in science. "Intellectual promiscuity" or "many facets of a personality" certainly describe quite well the result of a lifetime of seemingly diverse high-level activities. Remarkably, the achievements of Carl Djerassi appear in both arts, in the useful 'art' of science and in the art which is called 'pure.' Certainly his nature is rather that of a self-driven "Faustian" personality and not that of a fragmented character with several distinct and unrelated interests.

In recent years, Carl Djerassi has found his way back to Vienna, the city where he was born and spent his childhood and early youth. He has extended his previous commuting life between San Francisco and London to a triangle. Vienna is full of the spirit of baroque. Baroque art means complete artwork. The opera in Vienna stands very well for this idea: a luxurious building, singing and reciting actors and chorus, dramatic action, a splendid orchestra, beautiful robes, opulent scenery and finally the spectators, broadly engaged and interested in the performances. In fact, also the art of painting in the 20th century had a comparable evolution: In the mid of the century, the artist entered his own artwork when Lucio Fontana made cuts into the blank canvas. "Performance" became more and more an element also in the fine arts. It is no wonder that Vienna was one of the centers of the art movement of "actionism." In this town everybody can be an actor and spectator at the same time. Carl Djerassi's return to Vienna is good for Vienna and will hopefully be inspiring also for him.

A comprehensive definition of art includes the triumph of the human freedom of the imagination. Many artists spend their lives according to this paradigm. Scientists – in contrast – still live in reductionist worlds and tend to limit themselves to their specific scientific problems and fields. Djerassi has never reduced his personality to prescribed limits, but has covered the full scope of science – from taking notice to research to knowledge and finally to the art of implementation. He is certainly unique. He has always been at the forefront, a scientific trend barometer. Does this role also relate to his present activities, which seem so far from science? We hypothesize: Carl Djerassi might serve as template for the scientist in a future world. By demonstrating that one can live a life in which science is art and art is science, and in which art can be in everything, he deserves to be named "SciArtist."

This article includes material from Christian Noe, "Carl Djerassi – SCI-ART-IST," published in Pioniere der Sexualhormonforschung, *eds. R. Werner Soukup & Christian Noe, Wien: Ignaz-Lieben-Gesellschaft, 2010, 158-174.*

Works Cited

Djerassi, Carl. 1992. *Die Mutter der Pille. Eine Autobiographie.* Zürich: Haffmans.

Haberlandt, Edda. 2010. "Ludwig Haberlandt (1995-1932): Rückblick auf das Forscherleben eines Pioniers der hormonalen Kontrazeption." *Pioniere der Sexualhormonforschung.* Eds. R. Werner Soukup & Christian Noe. Wien: Ignaz-Lieben-Gesellschaft. 55-67

International Pharmaceutical Federation. 2012 (in press). *Impact of the Pharmaceutical Sciences on Health Care: A reflection over the past 50 years. A report written under the auspices of the Board of Pharmaceutical Sciences.*

Rao, Arun & Piero Scaruffi. 2012. "A History of Silicon Valley, The Largest Creation of Wealth in the History of the Planet." www.scaruffi.com/svhistory/sv.html

Science-in-Theatre

Balancing Act:
Drama about Science based on History

Robert Marc Friedman

Science, history, and drama: at first glance these categories suggest a rather shaky tripod of differing professional interests. Can drama offer insight into the scientific enterprise? Can theatre provide a forum for exploring the history of science? Can the demands of artistry and scholarship be harmonized? More to the point: By what standards, and by whose standards, might we evaluate a play, based on history, about science? Problems abound.

Why do playwrights bring history into the theatre? Why do some of them go through the motions of breathing life into historical figures and events on stage, only then to plea when confronted with misrepresentation, that in reality they are merely engaged in fiction? As artists, they claim, are we not free to interpret and portray at will? Why do historians get hot under the collar as soon as they see factual detail and contextual understanding trampled under foot? Why can they not sit back, relax, enjoy, and even have some fun, without fretting over the morality of historical distortion on stage? And why are some scientists unable to accept that others than themselves might have something of importance to say about the scientific enterprise, or even accept that they themselves might be lacking in critical perspective beyond self-serving platitudes? The world of science, and Western society more generally, tends to suffer from a state of living in a perpetual present; the past offers a resource for conceiving a future. What might be an appropriate ante of expertise in all three realms to play the game of creating science-theatre based on history? What chance do we have of creating a theatrical event that is honest to all three points of departure?

Carl Djerassi has with much dignity and passion brought insights from his extraordinary life in science to the stage and also into novels. He has worked with purely fictionalized stories informed by deep wisdom about science and its impact on society as well as with historical episodes molded into drama. He has touched the minds and hearts of audiences on several continents. That one of the world's leading researchers accepted the importance of following his call to educate, inform, and entertain colleagues and general public using the crafts of art is of course remarkable. He has most warmly received me as a comrade-in-arms in the use of drama to explore the world of science, and for that I am most grateful. Some of the following reflections were first published a decade ago, following upon the performance of my first play about science, based on history. Carl Djerassi read these with interest; he responded with much generosity. Much has transpired in the subsequent decade; my questions about our enterprise remains even as insight matures.

Playwrights have drawn upon history from the earliest times. Historical drama, in its many forms and variations, is of course a recognizable genre in the theatrical tradition. Yet history and drama are not necessarily compatible with respect to ends and means. True, both the historian and playwright need to get into the heads of the persons about whom they are writing. Still, from Aristotle to Pirandello and beyond, we know that truth does not have to be plausible, but fiction does. Playwrights reject historical facts for reasons of their implausibility – on stage the actual events, even with the most artful of foreshadowing, can at times seem too contrived. Historical villains might prove to be too melodramatic to serve as theatrically interesting antagonists. Instead, the constructed historical drama at times presents an artifact seemingly more plausible, and truer, than the actual historical record. But how far dare a dramatist deviate from that record – and of course whose authoritative record? What is at stake – why the fuss?

Even when the audience is well aware that the factual content of a history-based play is false, it may still find the play engrossing, urgent, and in touch with some deeper truth. Should we agree with those who claim that sometimes playwrights necessarily have to depart, and should depart, from fact in order to provide accurate and engaging portrayals of underlying historical forces far more important than biographical details? Yet, on closer inspection, as numerous scholars have observed and analyzed, most historical drama seems to depict not the social relations or political processes at the time of the play's subject matter, but actually those of the dramatist. Schiller, Shakespeare, Brecht and countless others used and reconfigured the past to address issues related to their own contemporary political and social realities. True, historians also tend to write about the past with the present and future, more or less, in mind, still they *necessarily* seek understanding from within the relevant historical context using categories derived from that particular time and place. What is gained and what is lost by riding roughshod over historical context? In the case of science, as so much else, we need to assess the contingent and the necessary in how and why things have developed as they have. The past is a foreign country, and for science it is not merely our understanding of the natural world that has changed over time, but so too the nature and internal culture of the scientific enterprise, including its moral and other legitimation with respect to society.

Why use the names and events of history when the dramatist's purpose is to illuminate a thesis or to provoke discussion? In his highly successful play *The Physicists* Friedrich Dürrenmatt freed himself from actual facts and personalities in order to create a drama that playfully and forcefully raises questions related to ethical responsibility for research. But – what if? Indeed – what if we choose to begin not with a fictive story, but with history. What if we find the content of history so compelling, that we simply must try to shape it, mould it into dramatic form?

In the case of science we are on terrain generally foreign to traditional historical drama. Playwrights in the past tended to choose as historical characters persons who already had achieved legendary status: kings, queens, heroes, and others with whom the public would be familiar. As such audiences could appreciate how the dramatist interpreted or altered received history or myth to make a point of contemporary importance. Few scientists have achieved such status. Most scientists are not legendary; they are scarcely known to the public or even to contemporary scientists. When the playwright breathes life into a name out of history and creates a dramatic character that is as new for the audience as any fictional character, should there not be some sense of responsibility for how that person – especially if from the recent past – is portrayed? During a symposium on Michael Frayn's *Copenhagen* in which some historians of science expressed concern over various portrayals of individuals and events, a producer from a well-respected theatre retorted that an artist is only responsible to art. Really? I have subsequently heard actors and directors repeat this claim as if a universal truth that transcends any context of origin and transmission over time. She added as well that nobody takes theatre to be more than theatre. Well maybe I am just too old fashioned and still believe that history is too valuable a resource for democratic civil society to fritter away. I find myself returning again and again to the question: just how malleable is historical scholarship in the forge of artistic imagination before intellectual and moral integrity, snaps?

Perhaps I am oversensitive. Three decades ago I used my doctoral research as the basis for a screenplay for Norwegian state television (NRK) about the so-called father of modern meteorology, Vilhelm Bjerknes. I set the action to April 1940, the German invasion of Norway. Along with the invading troops, Bjerknes' leading German disciple came, a meteorologist largely responsible for introducing the so-called Bergen methods of forecasting (based on weather fronts) into Germany. Ludwig Weichmann had almost lost his professorship in Leipzig because he did not join the Nazi party. But this did not stop the director of the television film from putting Weichmann, then an officer in the Wehrmacht, into an SS uniform.

When I saw the raw film rushes, I urged the sequences be re-shot. Those in charge insisted there was no need. "Nobody will notice." Of course they noticed. Viewers registered the image rather than the spoken words: what was to have been a complex moral story became simply black and white, good and evil. Moreover, when family members in Germany received a copy of the Norwegian TV-Guide showing still pictures, they indignantly wrote to NRK saying their father/grandfather was turning in his grave.

Screenplay writers frequently have little control over how their carefully crafted words and scenes are brought to life during production. I recall pulling hairs while I revised to allow nuance and subtlety to shine and dazzle – only to find afterwards key phrases mutilated, foreshadowing sentences cut, and unintended meaning fill the screen. A historical film that attempts to recreate the past with meticulous realism invites the viewer to believe that the dramatic events and personages are also dutifully recreated accurately. It seems especially odd after making a seeming contract with the viewer with respect to authenticity, then to introduce significant fictive developments, such as an illicit love affair, theft of a colleague's research, or membership in the Gestapo. Is it as morally problematic to deliberately distort or lie about the past in a theatrical drama with no semblance to naturalism, which shouts, flashes, and bounces about its business in a manner that reminds the audience that what is on stage is not a realistic recreation of the past?

If we choose theatre to diffuse insight from the history of science, we must accept the obvious: theatre is not suitable for presenting comprehensive detailed narrative history, and normally ought not to try. Still, no medium, of course, can better convey the immediacy of emotions. Drama may well reside in scientists' social, moral, and professional dilemmas. How they struggled for recognition, for resources, for arriving at new findings, and for gaining acceptance for these. What scientists do in their daily lives is both similar to the struggles, joys, and tragedies that others experience, but they also engage in an activity with its own specific goals, methods, and professional cultures. The payoff of bringing history of science successfully to stage (and screen), in part, might well be measured in the degree by which particular scientific events and persons can be transformed into public property. Significant chapters in science history have a right to enter our cultural heritage and provide a reservoir of insight to be drawn upon. As witnessed in Carl Djerassi's impressive list of novels and plays: science is too important a societal activity to insulate its practices, results, and consequences from public scrutiny and reflection. Carl, as myself, does not consider these endeavors to engage with drama as merely 'public understanding of science'; he appreciates the need to raise the consciousness of his own colleagues as well as other academics.

Although as late as 1999 I was still considering film as the medium for dramatizing my research in history of science; I turned at that time to theatre. At the first of two *Copenhagen* seminars in Copenhagen, arranged by the Dr. Finn Aaserud of the Niels Bohr Archives, I was able to thank Michael Frayn for reviving my belief in a theatre of ideas. Many decades earlier, I had abandoned thoughts of becoming a playwright when as an impressionable drama student in New York in the late 1960s, I heard during the intermission of a Harold Pinter play, the following comments: Wife: "Harry what did that mean?" Harry: "I dunno, but at 20 bucks an hour for parking, let's get the hell out of here." I repeat this anecdote not so much to reveal my earlier cowardice/prudence, but to keep in mind that a play must be able to attract, and hold, an audience. Most monographs in the history of science are written for a specialist public smaller than most theatrical audiences. What within the history of science might lend itself for creating a play that can attract a producer, money, director, and an audience? Of course history and science are only part of the equation; any number of dramatic and aesthetic traditions, technological and artistic resources, can help breathe life into historical episodes that might otherwise appear less than promising.

Let me turn to my own first efforts as one example. In 2001 I published a book entitled *The Politics of Excellence: Behind the Nobel Prize in Science*. This work offers a history of awarding the Nobel physics and chemistry prizes, as seen from the perspective of the Swedish committee members. I analyze the history of why and how individuals and groups attempted to use the Nobel prizes to further specific scientific, cultural, and personal agendas. The Nobel medallion is etched with human frailties, but can the noble and ignoble efforts of committee members to fulfill their difficult task be transformed into theatre? In a dramatization it would be foolish to deviate too radically from the historical analysis as the topic is highly charged and invites inspection of detail, especially as my research challenges well-entrenched beliefs about the Prize. Popularizing through drama should not mean simplifying – but rather engaging with a different type of craft with different conventions than that used when writing historical scholarship. I find the process of conceptualizing and writing a play to be an extension of my historical analysis, a process that can offer insight into riddles and ambiguities that many years of research still have not fully resolved.

Still, no matter how fascinating particular scientific episodes and personalities might be, these can induce snoring among theatre goers if the craft of playwriting is not respected. Herein lays the challenge. The historian can of course guide a reader through a thicket of detail. She can regulate the pace of narrative to allow meditation over a sticky interpretation, or take us by the hand on a detour from the main thesis in order to introduce background information, all without losing the reader. But whereas the historian and the novelist can tell a story, the playwright of course must show it.

If we consider the basics: Dramatic dialogue is created to set up conflict and action. Dialogue cannot remain static to ponder a detail. An audience tends to choke when served large indigestible servings of background information (indeed such a play no doubt will not reach a stage). The protagonist must want something strongly; and other characters who are also motivated by strong desires provide resistance, thereby propelling the drama forward. Secondary characters must be introduced on the basis of the dictates of plot and dramatic logic rather than on the need for historical completeness. Yes, professor X may well have had help from her trusty assistants all the time in the lab, but just what sort of work might these secondary characters be doing in and for the play? The dramatist writes in the present tense; and even when a character confronts ghosts of the past, on stage it normally must happen in the theatrical present. Such differences in means and goals for exposition need not be insurmountable, but they require respect.

To use drama as a means to educate and entertain, to provide an intellectual and emotional event, obviously requires adopting a bit of modesty towards the task. I decided to begin with one-act plays. I planned that one of these would focus on why the Nobel committee refused to acknowledge physicist Lise Meitner's contributions to the discovery of nuclear fission. I had not come very far when Swedish actors who had performed *Copenhagen* and who were in attendance at the 2001 symposium in Copenhagen asked whether I would be interested in writing a one-act play on Meitner for the Gothenburg International Science Festival. I accepted.

Lise Meitner's life and career in physics can surely be examined, explored, and shared with a broad public though a dramatized film; and can also offer materials for crafting a full-evening theatrical event. But what can be achieved in the format of a one-act play? A one acter normally follows a rigid unity of time, space, and action: if a scene change is to be included, it must be achieved with minimal interference on stage. A one-act play should open near the point – or even just after the point – when some incident sets in motion the events that bring about the play's action. In a drama of, say, fifty minutes, very few characters can be introduced and generally only one of them can be developed.

With Frayn's *Copenhagen* and Pirandello's *Six Characters in Search of an Author* teasing my imagination, I considered the possibilities of the historian interacting with the characters. If historical research permits questioning the manner by which some scientists have been granted a place in the Nobel pantheon and others denied – then this very process of liberating history from myth and misconception might offer a framework for a play. Scientists from the past in the form of dramatic characters might confront myth, hagiography, and selective memory and in the process might allow recapturing a greater fullness as humans while also giving voice to the passions and practice of life in science. Well, that at least was the thought with which I began.

The one-act play, *Remembering Miss Meitner*, is set in the theatrical present. Recent historical scholarship bring the figures of Lise Meitner, Otto Hahn, and Manne Siegbahn back to life, where they confront the historians' revelations. In early versions of the play the two male scientists, both of whom are sleeping in the cocoons of accepted reputation as Nobel laureates when they are forced out into the light, are forced to accept re-evaluation and react brusquely.

The cause of the play's action is clear from the opening: Meitner is reading the new historical literature, such as Ruth Sime's excellent biography and my own study of the Nobel Prizes. She is gaining strength to confront those whom she feels treated her poorly. Hahn and Siegbahn are upset over the new historical interpretations and threaten to refuse to perform in the intended play. Indeed, in more recent productions, I revised the text so that the characters are actually summoned to a staged-reading of a full-length play (which I happen to have in preparation on the back-burner), and while waiting for other characters to arrive, they begin to discuss what Siegbahn and Hahn consider a scandalous play. Their increasingly heated exchanges about the play in which they are to perform brings them into confrontation about the past and the newly revealed secrets.

The primary 'stuff' for the play then is a revisionist historical agenda. Reward in science, like elsewhere in society, is far from perfect. Lise Meitner devoted herself to physics. She understood that the world of science in the first half of the twentieth century had its share of social imperfections, but that did not stop her from embarking on what became a significant career in physics. Nazi persecution stripped her of possessions and employment; in 1938 she fled to Sweden. Three injustices that followed upon this misfortune brought further grief as they shattered her self-confidence and reputation. Having been the initiator and scientific leader of the team that included chemists Hahn and Fritz Strassmann, and even secretly remaining in contact with Hahn after fleeing Berlin, she was denied credit in the discovery of nuclear fission. Even after the war, when it was no longer necessary to keep their continued collaboration secret, Hahn refused to remember her role during the crucial months after she had left, and also conveniently forgot his own confusion and misconceptions while initially allegedly 'discovering' fission. In Stockholm as a sixty-year-old refugee and internationally prominent nuclear physicist, Meitner was, to her mind, shabbily treated and hindered from continuing her research by Manne Siegbahn, who guarded his authority and resources jealously. Moreover, the Nobel committees for both physics and chemistry clearly chose to ignore Meitner's contributions to the discovery and explanation of fission. In 1945 Hahn alone received the previously reserved 1944 Nobel chemistry prize; the physicists refused to acknowledge fission with a prize. Siegbahn and other committee members relied on the secrecy of Nobel proceedings to bury their biased and faulty evaluations. Meitner remained silent. History now suggests how and why these events happened.

About two hundred people were in attendance at the first performance of the play in April 2002 at the international science festival in Gothenburg. When near the end of the play, Lise Meitner movingly relates the rapture of research as she recalls her walk in the snow when she and her nephew Otto Frisch finally made sense of nuclear fission, goose bumps broke out on my arms, tears filled my eyes – and I knew it was all worth the effort. We surely had the audience in our grasp! Only about seven more minutes of dramatic concluding events ... Finally the stage darkened; enthusiastic applauds erupted. Comments from the audience confirmed my growing optimism. But – when the initial favorable response settled down to allow for a panel discussion, other tones emerged.

The panel discussion revealed that the play was most certainly capable of provoking debate. Defensiveness over the Nobel institution and Manne Siegbahn came to the fore. One member of the Nobel establishment, initially had hailed the play as brilliant, but then tried to find less controversial explanations for why Meitner did not receive a share of a Nobel Prize. Claiming that it was not usual to divide prizes at the time, he chose to ignore a multitude of facts and continued to use his own reputation as a physicist connected with the Nobel establishment to criticize the play for being too controversial. Suddenly the physicist was both historical expert and theatre critic. One of Siegbahn's disciples first praised the play as drama but then insisted it contained a number of 'lies'. He insisted that feminists and Zionists were behind the efforts to portray Meitner as a heroic martyr. He subsequently attempted repeatedly to discredit both the play and the historical research upon which it is based, while revealing minimal insight into the nature of historical scholarship.

True, the scope of a one-act play does not allow for sufficient nuance in the portrayal of all the characters. The goal was not to paint Siegbahn black, Meitner white, and Hahn yellow with cowardice. Both men offer key lines that provide insight into their actions and why they should not be portrayed monochromatically. The three excellent actors, Inger Heymann (Meitner), Ingemar Carlehed (Hahn), and Johan Karlberg (Siegbahn), maintain the ambiguities and complexities in motive and intention in their interpretation so as to avoid melodrama. Meitner's reactions and accusations do imply some murky and puzzling personality traits of her own that also need further investigation in the scope of a full-length play.

More problematic is the question of whose history shall serve as the basis for drama. Scientists frequently consider history as an arena for heroic tales, a field of honor that allocates prestige to the worthy. As we know, it is not unusual for disciples of a significant researcher to assume the role of historical gatekeepers.

But history can be judged – good or bad, solid or sloppy – by a wide range of accepted professional criteria and craft skills. Truths may not be black and white, fixed for all time; absolute objectivity and positivistic nuggets of historical truths may be a belief of the past, but our understanding of the past is not 'anything goes.' Debate and disagreement among scholars advances our insight, scholarly publications are normally part of a broader discourse. If a non-specialist wants to make use of a particular book either in a drama or to argue against a drama, then the reception of that book in the relevant scholarly community is an essential prerequisite. Liking and not liking a work of scholarship simply because it confirms or upsets self-understanding, legend, or reputation is quite a different matter. A true friend of science should not deliver sanitized stories; history written with a Hollywood "feel good" coating.

In this respect we might consider whether science-theatre can flourish as something more than public relations? Do research and arts councils support voices that raise hard-hitting critical questions? Following the success of Michael Frayn's *Copenhagen* it seems that science has become a relatively popular topic for playwrights. I cannot admit to having followed developments closely during the past decade, but a number of indications make me wonder whether the recent alleged mutual embrace of science and the arts may not necessarily always be something so positive. Do some of these activities lead to a trivialization of science and the history of science? Is there too much mutual admiration at the expense of critical thought? Do scientists want to use the arts for public relations and as proof that science, too, belongs to culture (which it does), or are they willing to accept that they, just as the rest of society, also need to be scrutinized? Has science become just another fashion for momentary exploitation by artists, some of whom glibly delight in a wealth of metaphor and imagery, while remaining ignorant of any deeper understanding of the subject? Just what should we aim for when considering interactions between art and science? Does the use of scientific ideas or technical jargon as metaphor in a novel or play constitute a bridging of the so-called two cultures? Of course there are many more than two cultures: the arts as practiced are not one and the same as the arts as analyzed in academic departments of literature, art, theatre, and cultural studies. In the academic humanities momentary intellectual fashion and theoretical mass-movements tend to favor prioritizing bold originality of form; the use of theatrical and artistic experiment as fodder in the academic game of careerist posturing. Excellence in traditional craft skills and razor-sharp, imaginative understanding of characters may not be cutting-edge in some academic fields, but can appeal to producers, directors, and theatrical publics. Bridging the 'two cultures' between the sciences and academic humanities programs is one thing, bridging with the performing arts is often rather something else.

Who are the gatekeepers of the various funding agencies that offer stipends for writing new science-related plays? Are the Alfred P. Sloan Foundation and comparable private and public grant-awarding institutions open to critical voices? To some it appears that the Royal Society's past public understanding of science campaign, and surely many other comparable endeavors, tends to weed out perspectives that dare portray the world of science as anything less than rosy and inviting. And yet, many scientists, including important leaders, are well aware of the need for critical reflection and self-examination: the frightening lack of interest in science among young people may not simply be the result of anti-scientific propaganda coming from arts, humanities, and social sciences.

The subsequent history of *Remembering Miss Meitner* reinforces my belief that theatrical events based on professional historical scholarship can offer opportunities for constructive dialogue among scientists, the general public and historians. The play has subsequently been on the repertoire of the Gothenburg City Theater [*Göteborgs Stadsteater*] for seven seasons as part of the popular Lunch Theatre, a forum for one-act plays. General audiences responded with enthusiasm; frequently most of the public remained up to thirty minutes and more to discuss the play with the actors (and playwright, when present). Swedish national broadcasting adapted the play for its weekly Radio Theatre and broadcast it six times over recent years. Although some members of the Nobel establishment and Siegbahn school continue to disparage the play and the historical scholarship upon which it is based, it was nevertheless invited to the Nobel Museum in Stockholm. Some scientists have embraced *Remembering Miss Meitner* with pure enthusiasm. In particular Professor Björn Jonson, now retired from Chalmers University of Technology and one of Sweden's leading physicists, who had originally initiated the idea of a Meitner play and supported its development, organized performances of the play, in English, at the 2004 International Nuclear and Particle Physics Congress in Gothenburg. On the main stage of the theatre, large full-house audiences responded with enthusiastic standing ovations. Another prominent scientist, Professor Francesco Iachello of Yale University, was so moved by what he experienced at that congress, that four years later he arranged for the same actors to perform the play at Yale University, where students, including those attending a national conference of undergraduate women physics majors, and professors of various disciplinary backgrounds responded unreservedly. Comparable guest performances have been generously well received around Sweden, Oslo, and at the Maxim Gorki Theatre in Berlin. Again, especially in Berlin, long and emotional discussions followed the performances. Other highly satisfying responses included staged-readings at the Graduate Center at City University of New York and Oak Ridge National Laboratory in Tennessee, where researchers and residents took the play to heart, giving it almost cult-status after numerous performances at the local theatre and museum for science and technology. An Italian production in Bologna left the university's rector in tears. Most importantly wherever the play has been staged, it prompted discussion and raised both scientists and non-scientists awareness of discrimination in science as well as the importance of other values in science than winning prizes and striving for academic eminence.

I am currently working on a non-naturalistic, highly theatrical play about polar heroes Fridtjof Nansen and Roald Amundsen that draws upon recent historical research while challenging popular conceptions of the two, especially with respect to the role of science in their endeavors and their relationship. As I find myself standing again on that rickety three-legged platform, I hope it holds, at least for yet another attempt at balancing historical, scientific, and theatrical sensibilities. World premiere will be at Hålogaland Theatre in Tromsø as part of the national celebration of Amundsen, Nansen, and the Norwegian polar heritage. I am sure Carl would approve of my efforts to inform director, actors, and dramaturg that regardless of the degree of theatrical fantasy written into the play, we must get the science right and stay within the realm of translating historical scholarship into responsible art. I so much appreciate his good-will and valuable encouragement over these past years.

In this contribution, I have in part drawn upon materials presented in *"Remembering Miss Meitner*: an attempt to forge history into drama," *Interdisciplinary Science Reviews*, 27 (2002), 202-210, but have updated it based on subsequent experience with this and other plays.

"Ambition without Love is Cold": Priority and Kudos in *Oxygen* by Carl Djerassi and Roald Hoffmann

Eva-Sabine Zehelein

> *And sometimes the heart comes into it.*
> *(Stephenson, Experiment 47)*

One might take it as a given that ambition is a wide-spread human character trait, and that a substantial amount of ambition characterizes every successful scientist. There are a number of theater plays such as *Oxygen* (2001) by Carl Djerassi and Roald Hoffmann, Shelagh Stephenson's *An Experiment with an Air Pump* (1999), or Vern Thiessen's *Einstein's Gift* (2003), which focus primarily on the emotions and inner motivations of scientists. The playwrights ask not only what effect unbridled ambition can have for a scientist and his work or career, but also how it affects society. After all, the 'tribal culture' does not exist in isolation; the scientists' actions have reverberations for all of us, introduce ethical quandaries, political concerns and legal ramifications.

Priority, power, and fame are identified as the driving forces of the natural scientists in *Oxygen*, which centers around the question 'what is discovery?' and the historical dispute over 'who discovered oxygen?' Was it Lavoisier, Priestley or Scheele? And what role did the respective wives and partners play in the discovery? *An Experiment with an Air Pump* presents different facets of the scientist. In 1799, in the thrill of Enlightenment thought, the analytical and the more 'humane' scientist clash in a story of betrayal and lost hope. The physician Fenwick believes that the heart has to come into the focus of science, whereas his colleague Armstrong is cold of heart. He toys with servant Isobel's emotions and calculatingly snuffs her out so that he can analyze her bodily deformity. And in the parallel storyline, in 1999, Ellen decides to continue her career in genetic engineering because of the fascination and passion inherent in the scientific endeavor and swallows the ethical scruples voiced by the Geordie builder Phil and her husband, the unemployed English professor. *Oxygen* and *Experiment* both use two different time lines for their respective plots, aiming at a comparison of 'then' and 'now.'

In *Oxygen*, the internal maneuverings of the 2001 Nobel Committee mirror the wheeling and dealing of 18th century scientists and their women, betraying the naïve belief in the purity of the scientific endeavor. In *Experiment*, the same problems are debated in 1799 and 1999: the nature and future of progress, the conflicting *conditio humana* in view of man's determination to 'decode' nature, and the resulting ethical tribulations. *An Experiment with an Air Pump* illustrates the ethical consequences of progress, exemplified through the field of biology, and here the 'hot topic' genetic screening and engineering. At the core of the very complex symbolic net resides the dichotomy between moral ethics and scientific rationality which has remained a constant in our lives ever since. *Einstein's Gift* also brings to the stage the ethical responsibilities scientists have to face by unmasking in a different context, namely Fritz Haber, Albert Einstein and the two World Wars, that science is never an 'uncontaminated' endeavor. Thiessen elaborates on the question what happens when science becomes political and when scientific research is (ab)used for military purposes, in other words, when science discovers sin. And all three dramatic works use the tribal community of the scientists as an exemplary group to illustrate that "ambition without love is cold." (*Oxygen* 104) This contribution will highlight the special case of *Oxygen*.

Oxygen, the play written by "Carl Hoffmann and Roald Djerassi,"[1] premiered in San Diego in April 2001. Most certainly one is justified to wonder why anyone would ever write a play about a – at ordinary temperatures – colorless, odorless and tasteless gas, slightly heavier than air. This play is not primarily about oxygen; rather, the plot deals with the fundamental question 'what is discovery' exemplified by 'who discovered oxygen?' This is also a play about 'women's Buts' (in Scene Three, the Scheele and Lavoisier dialogue counts eleven "buts" on less than three pages) and here too, we find the overarching Djerassi-topics, the four 'Ws': women, women in science, women with good legs, and the regal 'we', including or connoting the scientists' strife for priority, power and fame. *Oxygen* highlights the driving forces of the scientific tribe in a special setting, based on a historically verified dispute.

The action alternates between 1777, when the three contestants for the title "Discoverer of Oxygen" and their wives meet in Stockholm at the invitation of King Gustav III, and the year 2001, when a Nobel committee of four (plus one) is meant to award the first "retro-Nobel" for path-breaking scientific achievements made prior to 1901, the year the Nobel Prizes were established. They opt for the

---

[1] Cf. Djerassi in an interview together with Hoffmann. Both stress that the play was a joint venture, an absolute collaboration, ultimately creating, as Djerassi suggests, one authorial persona, "Carl Hoffmann and Roald Djerassi." (Devins, 18. See also Hemming.)

discoverer of oxygen, but then wrestle to decide on the person to go with this discovery: Carl Wilhelm Scheele, Joseph Priestley, or Antoine Laurent Lavoisier.

The Swedish apothecary Carl Wilhelm Scheele (1742-1786) was the first to isolate oxygen, or "fire air," but he did not place it in the accurate framework by clinging to the then current "Grand Unified Theory" of phlogiston which contests that all flammable materials contain an odorless, colorless, weightless substance (phlogiston) that escapes upon burning, and Scheele failed to publish his results until 1777. Joseph Priestley (1733-1804), the British Unitarian minister and chemist, too, identified oxygen, and also still adhered to the wrong theory of phlogiston. Yet he published his experiments first and called the gas 'dephlogisticated air.' And Antoine Laurent Lavoisier (1743-1794) was the first to fully understand the nature of oxygen, to name it, and to advocate and promote this new discovery as the scientific revolution overthrowing the old system of phlogiston by recognizing that combustion involves oxidation. Lavoisier also introduced the concept of balanced equations in chemical reactions. (Balaram 925f.)

The play gyrates around the core issue: 'what is discovery?' and embeds the question into an illustration of the internal mechanisms which determine the scientific community. Is "discovery" defined as the first find, the first publication, or the first full understanding? A question one could try and tackle theoretically or philosophically by taking Thomas Kuhn as a stepping stone and continuing to post-Kuhnian debates from there:

... discoveries ... are not isolated events but extended episodes with a regularly recurrent structure. Discovery commences with the awareness of anomaly, i.e., with the recognition that nature has somehow violated the paradigm-induced expectations that govern normal science. It then continues with a more or less extended exploration of the area of anomaly. And it closes only when the paradigm theory has been adjusted so that the anomalous has become the expected. (Kuhn 52f.)

Parallels between Kuhn's argument and the play's structure could be unraveled, eventually leading to the insight that the play *Oxygen* can be considered a fictional illustration of Kuhnian rhetoric and theory. Yet that would carry way off the authors' mark, and the play does provide a 'sort of answer', since it is open-ended. The committee decides to vote not for a single scientist, but for a pair. Astrid and Bengt both favor Lavoisier, but we do not learn who their respective second choice(s) will be. Sune (advocating Scheele) and Ulf (representing Priestley) discuss voting for their couple Scheele and Priestley to prevent the nod to Lavoisier. The final results though are not disclosed.

Hoffmann and Djerassi could have broadened the scope of their play by including references or even sub-plots illustrating the thorny times the three men were witnessing; the late 18th century was not only a time of scientific, but also of social revolution and turmoil. Lavoisier, the scientific revolutionary and conservative royal tax collector, was guillotined on 8 May 1794 at the Place de la Révolution during the Reign of Terror that ruled Paris. Priestley, the scientific conservative, who would not discard the theory of phlogiston, sympathized wholeheartedly with the French Revolution. In July 1791, he organized a festivity at his Birmingham home to commemorate the fall of the Bastille, and that day, his home, laboratory and library were destroyed by royalists and Anglican Church zealots. In reaction, Lavoisier wrote to Priestley:

As a citizen, you belong to England and it is for her to atone for your losses; as a Scholar and as a Philosopher you belong to the entire world; you belong above all to those who know how to appreciate you, and it is we, united in agreement, who vow to restore to you the instruments which you have employed so usefully in our instruction. We have therefore resolved to re-establish your Cabinet, to raise again the Temple which ignorance, barbarity and superstition have dared profane. What more important service can we render to science than to place in your hands the instruments necessary for its cultivation? (Balaram 925f.)

These and many more fascinating historical nuggets could have been employed. However, their inclusion would have shifted the play's emphasis in a completely different direction, and they are thus still awaiting dramatization.

The time shift between 1777 and 2001 nicely illustrates the topics under discussion, namely that priority, recognition and fame have remained unscathed over the centuries and that in contrast to commonly held naïve notions science was actually never a pure idealistic and unpolluted endeavor. The wheeler dealers of the play are not the (three) eminent scientists, but the women, in 1777 just as much as in 2001. *Oxygen* is clearly a women's world. Intriguingly, www.oxygen.com is the first online and on-air network from women for women, with the slogan: "Another Great Reason to Be a Woman." Yet the emancipated and initiated author wonders whether series such as "Talking Sex with Sue Johanson" really contribute to the women's endeavor when they deal with questions such as "Does swallowing sperm cure cramps?"

The first scene provides a classical, not to say Aristotelian exposition of both characters and themes. In 1777, the three 'wives' (Fru Pohl is not yet married to Apothecary Scheele) converse in a Stockholm sauna. As Scheele remarks: "Not much is hidden in a sauna" (83) – an aspect the authors really hit home through repetition: "Lavoisier: Nudity can be disarming" (39), and "Mme. Lavoisier: Some things ought to be hidden ... even in a sauna." (40)

It is in the course of the ladies' conversation in the sauna that the key topics of the entire play are presented *in nuce*: all three parade their partners with whom they identify and explain their roles in their respective marriages or relations. Madame Lavoisier introduces two important concepts characteristic of the scientific tribe, which is strife for reputation, connected to the royal 'we.' "Mme. Lavoisier: ... my husband told me something very useful. 'The product of science is knowledge ... but the product of scientists is reputation.'" (5) And Fru Pohl mentions the ominous letter Scheele wrote to Lavoisier three years before, in 1774, which will turn into one of the leitmotifs in the course of the action. A wonderful character exposition appears right in the first lines:

Mme. Lavoisier *(dreamily)*: I have never been beaten before... not like that. Can we do it again?
Mrs. Priestley: Madame! In England the birch is used for chastisement.
Fru Pohl: In Sweden, we consider it healthy. It brings the blood to the surface. So much better than leeches. (3)

In this exchange, which becomes ever more heated, Mary Priestley, aged thirty-five, is depicted as a prim, prudish, proud and religious wife and mother of four. Fru Pohl, aged twenty-six, is the practical, down-to-earth woman in this trio, a young widow with a son, who now keeps house and maybe everything else for Scheele. She is committed to Apothecary Scheele by shy love.[2] The focal point and counterpart to Fru Pohl is lascivious and lavish Madame Lavoisier, sweet nineteen. Sensuous and passionate, yet at the same time dedicated, not to say devoted to her husband's science and his close collaborator, she identifies completely with him and his work and uses the royal 'we' rather excessively:

Mrs. Priestley: What do you mean?
Mme. Lavoisier: We are not convinced –
Mrs. Priestley: *We*?
Mme. Lavoisier: My husband is not convinced... and therefore, I am not convinced.
(6; emphasis in original)

All three women pay a price for their husbands' 'success.' The mechanisms of the respective partnerships are illustrated in Scene Seven with three flashbacks to the year 1774, which are embedded in the discussions of the 2001 committee. Fru Pohl suffers in silent love and admiration for her "chemical monk" as the 21st century scientist Sune Kallstenius calls him. Programmatic example:

---

[2] In his review of the San Diego production, Steven Oxman has accused the play of its "amateurish clunkiness," and provides two examples, one of which is Fru Pohl's sentence to Scheele: "I would help you, Carl Wilhelm. If only I were not so ignorant..." (p. 48 of the play). Oxman's ensuing comment clearly reveals his own ignorance of diversified character delineation: "the scientists/playwrights are not exactly presenting themselves as capable of crafting women characters despite their every effort." That Oxman misspells "Hoffman" is a minor blunder compared to his failure to find intellectual access to the dramatic plot and themes. Thus, it is perfectly understandable, maybe even forgivable, that for someone like Oxman *Oxygen* appears a "rather labored play" which "just never delivers dramatic excitement." (Oxman, "*Oxygen*")

Fru Pohl: ... Carl Wilhelm ... I worry about your health.
Scheele *(Moved, takes her hand, pauses to inspect his hand and then hers)*: Look! The coffee sticking to your hand! Is it some form of magnetism? (49)

Mrs. Priestley can feel love within the set limits of her religious belief and her corset of primness, and Mme. Lavoisier commits herself to her husband's success trying to thus sublimate his inhibited love and compassion for her. Mme. Lavoisier is the play's most well-rounded character. She is a multi-functional weapon, combining femininity with traditionally male attributes such as strife, cunning, and determination. She demands equal partnership in the scientific endeavor. Madame Lavoisier is the embodiment of a successful 'bridging the gap' by combining the dry factuality of the sciences (she takes down the numbers, makes lists of experiments) with the aesthetic (she draws and etches plates). "There was chemistry to study. Art too ... all to help my husband" (6), "I helped Antoine in the laboratory ... as in the salon." (9) Yet she complains that although she assists him in ever so many ways, he never discusses the serious theories or findings with her but with "men," colleagues. (9) She is thus being reduced to a handmaid, and not recognized as an equal partner. Her remark to the audience in the first Intermezzo, which is dedicated exclusively to her monologue, is programmatic: "So ... we talk women's talk. ... Wearing the woman's mask ... her husband's face on it ... smiling politely." (9) We find similar characters in other plays, one outstanding example being Michael Frayn's *Copenhagen*. Here, Margrethe Bohr is the pivot of the entire action. She serves her husband's and Heisenberg's science by typing the manuscripts, by acting as the sounding board for their hypotheses, and she negotiates between them. Yet she, too, is not appreciated in her triple role as scientific collaborator for menial tasks plus wife plus mother. As far as her functions in the play are concerned, her role is even more complex, since she also represents 'the public' for whom the scientists explain in plain words their complex theories and finds.

In *Oxygen*, for the sake of success and priority, Madame Lavoisier takes matters into her own hands in the style of Felix Frankenthaler in Djerassi's "ICSI" play *An Immaculate Misconception*. In 1774, she withholds Scheele's letter from her husband, and the same year she filters information by only selectively translating Priestley's new observations to her husband at a Paris dinner. (41) And she seeks a lover's fulfillment in another's arms – much later in the play we learn of her seventeen-year long affair with Pierre Samuel DuPont, the French father of the American chemist-turned-millionaire DuPont. At the nerve center of the action, she tried to bend the course of history. Scene Ten lifts the secret: Ulla Zorn, the science historian on the Nobel Committee, the 'plus one-member,' exhibits the exciting result of her hands-on research: she displays slides of a pseudo-book, a *nécessaire* disguised as a book aptly titled *Histoire des Théâtre* [sic!]. (95)

The audience sees various photos of a travel chest consisting of a tray with many compartments in one half; underneath the tray there is room for stationery, and a broken mirror in the other lid. The contents once more illustrate Mme. Lavoisier's multi-facetted and complex character. We find thread, needles and scissors as the tools of the practical (house)wife. Although the scissors are not mentioned in the play, they are visible on the slides reproduced in the book version and projected on the stage. Then there are combs, the mirror and perfumes – typical female *accoutrements*. And we also discover pens, ink, stationery and a ruler – writing and measuring pointing towards her role as chronicler and active participant in her husband's scientific research. The ruler, "crammed in a slit, like in a Swiss Army knife" (97), elegantly illustrates the science wars, wrapped up in a book on the history of theater. "Science-in-theatre" lives up to its most literal meaning. And behind the broken mirror of that travel chest, Ulla Zorn has discovered a letter Madame Lavoisier wrote to her husband during his imprisonment in December 1793, roughly five months before his execution. Here, she explains and tries to justify, more to herself than to her husband, why, years back, in 1774, she had intercepted Scheele's letter to Lavoisier:

Now that the brilliance and accuracy of your studies have convinced the world of the central role of oxygen in chemistry, now that phlogiston lies in the dustbin of discarded theories ... I will not speak of the diehards such as Dr. Priestley who continue to preach it. *(Pause).* I ask you now to forgive me. I could not show Apothecary Scheele's letter to you, my dear husband. It would have taken the wind out of your sails, you, who were so close ... And I told you why I felt incapable of destroying it. Our priority rested on my hiding it. (99)

"Our priority rested on my hiding it." Again, she identifies with her husband's science; drawn into the maelstrom of priority, fame and recognition at all cost, she deliberately manipulated the course of history. Madame Lavoisier never sent the letter; the political situation had become too dangerous for such a confession, and Lavoisier was guillotined in May 1794. "One letter she could not send ... another she could not burn." (101)

A travel chest, attesting to the variety of duties the scientists' wives perform(ed), hidden in the cover of a volume on theater history; a play within the play, a masque, performed by the Lavoisiers at the court of King Gustav III, harking back to an actual play, now lost, the Lavoisiers staged for their friends and patrons; the fictional play "The Victory of Vital Air over Phlogiston," illustrating as a didactic mini-lecture in verse the paradigm change from the theory of phlogiston to the acceptance of oxygen; a setting, King Gustav's court, reminiscent of Verdi's opera *Un ballo in maschera*; and names for a 2001 Nobel committee where four out of five allow references to the arts[3] – discreetly and delicately, in a wonderfully multifaceted manner, the Two Cultures are bridged on various symbolic levels. Furthermore, the inclusion of the character Ulla Zorn adds a number of spicy interpretive doses to the play. Within the 2001 committee, she serves to illustrate a double confrontation: male–female, and chemistry–history. The male members of the committee voice highly pejorative opinions about history and historians:

Astrid Rosenqvist: What's wrong with historians?
Sune Kallstenius: It's a thing scientists do when they can't do science anymore.
Astrid Rosenqvist: But professional historians?
Bengt Hjalmarsson: What would they know about science? *(Pause)*. You might as well search the web! (16f.)

The chemists' attitudes reflect both a condescending view and a deep gap; history is nothing but the random accumulation of facts, deplete of specific knowledge or expertise. Ulf Swanholm elaborates in Scene Eleven:

... You're ignoring two fundamental differences between literature and science. The literati don't worry about priority. ... If Shakespeare had never lived, "King Lear" could never have been written. ... Consider oxygen. If Scheele or Priestley or Lavoisier had never lived, somebody would have discovered oxygen. The same with Newton and gravity, with Mendel and genetics – (108f.)

Priority, fame, and recognition are thus part and parcel of the scientific endeavor, and characterize, not to say determine, the scientist in stark opposition to the 'literati,' who produce a singular, unique and personal piece of art. As Madame Lavoisier has learnt from her revered *mari*: "The product of science is knowledge... but the product of scientists is recognition" (5) – in 1774, 1777, 2001, and, most certainly, also today. Ulla Zorn and Astrid Rosenqvist represent the 21st century professional woman, aiming to enter and endure within the male dominated realm of science. Astrid, just as Madame Lavoisier, illustrates that woman

---

[3] The names of the 2001 Nobel Committee members seem to be another intellectual detective game the authors are playing with their audience: Bengt Hjalmarsson, Sune Kallstenius, Ulf Svanholm, Astrid Rosenqvist. Kallstenius and Svanholm are famous Swedish artists, dead and alive. Edvin Kallstenius was a Swedish composer (1881-1967), Set Svanholm a tenor and one-time director of the Stockholm Royal Opera; and Barbro Hjalmarsson is a contemporary painter (Asiatic aquarelle techniques on silk). Astrid Rosenqvist, the committee's chair, has an alter ego in Helena Hillar Rosenqvist, MP of the Swedish Parliament for Östergötland County. In fact, within the play it is Astrid who operates akin to a politician, trying to negotiate a compromise between the other members. Yet probably the authors were thinking of Astrid Gräslund, the only current (and probably also first) woman member of the real Nobel Committee on Chemistry. And the science historian Ulla Zorn might allude to the very famous Swedish impressionist Anders Leonard Zorn (1860-1920).

has to make choices: children or career. Mme. Lavoisier's frustration surfaces in her monologue: *"(Mimics Fru Pohl's voice and intonation)* 'And no children?' *(Resumes normal voice and accent)* What gives Fru Pohl the right to ask? ... Not even married to Apothecary Scheele! *(Pause)* I helped Antoine in the laboratory ... as in the salon." (9) And in the 2001 version: "Ulla Zorn: Sorry about that. I just wanted to know what price you're willing to pay to be successful as a scientist ... and as a woman. Astrid Rosenqvist: I have no children. Many would consider that a heavy price." (35) Finally, Ulla Zorn is not an amanuensis, but a *nemesis*; yet she is also Clio, the muse of history and historiography who discovers the "book" *"Histoire des Théatre"* [sic!].

The play ends on the word "Imagine!" – once more tying up the strings of science, history, theater and fiction for a complete and convincing piece of "science-in-theater," which on its most subtle level, resonates one sentence: "Ambition without love is cold." (104)

Works Cited

Balaram, Padmanabhan. 2002. "Oxygen, Lavoisier and Revolution." *Current Science* 83.8: 925f.
Cohen, I. Bernard. 1980. *The Newtonian Revolution*. Cambridge: Cambridge UP.
Devins, Dorian. 2001. "Discovering Oxygen." *The Dramatist*. 4/2: 16-28.
Djerassi, Carl & Roald Hoffmann. 2001. *Oxygen*. Weinheim/New York: Wiley-VCH.
Donovan, Arthur. 1993. *Antoine Lavoisier. Science, Administration and Revolution*. Cambridge: Cambridge UP.
Frayn, Michael. 1998. *Copenhagen*. London: Methuen.
Hemming, Sarah. 2001. "Something in the air." *Financial Times* 10 November. www.djerassi.com/oxygen6/index.html
Kuhn, Thomas. 1996. *The Structure of Scientific Revolutions*. Chicago: U of Chicago P.
Martin, Colin. 2001. "Three Chemists in Search of a Gas." *The Lancet*. 358/9294, 17 November: 1735.
Oxman, Steven. 2001. "Oxygen." 5 April 2001. www.variety.com/review/VE1117797720.
Poirier, Jean-Pierre. 1998. *Lavoisier: Chemist, Biologist, Economist*. Philadelphia: U of Pennsylvania P.
Rivers, Isabel & David L. Wykes, eds. 2008. *Joseph Priestley, Scientist, Philosopher, and Theologian*. Oxford: Oxford UP.
Schofield, Robert E. 2000. *The Enlightenment of Joseph Priestley. A Study of His Life and Works From 1733 To 1773*. University Park: Pennsylvania State UP.
Stephenson, Shelagh. 1999. *An Experiment With an Air Pump*. London: Methuen.
Thiessen, Vern. 2003. *Einstein's Gift*. Toronto: Playwrights Canada.

Carl Djerassi:
U.S.-amerikanischer Dramatiker Wiener Herkunft auf dem Weg nach Europa

Isabella Gregor

Theaterstücke von einem Autor, der in jungen Jahren aus Europa emigriert ist und den Großteil seines Lebens als ‚nüchterner' Wissenschaftler in Amerika verbracht hat? Wie geht das zusammen? Nicht überraschend die Emigration, zu der so viele in der dunkelsten Zeit Europas gezwungen waren. Aber der Wissenschaftler, Dr. hc. mult. – davon eines, das Dortmunder Ehrendoktorat 2009 für sein literarisches Werk – widmet seine Arbeitskraft seit Jahren der Kunst: als Bildersammler, als Förderer unterschiedlichster Kunstprojekte mit Stipendiaten auf seiner Farm in Woodside/California, als Romancier und in den letzten Jahren als Bühnenautor.

Seine amerikanische Frau Diane Middlebrook inspirierte ihn, sich der Belletristik zuzuwenden und, von der europäischen Kultur – back to the roots – inspiriert, Theaterstücke zu schreiben. In England war es, so erzählt er, als Carl Djerassi mit seiner Frau in London im Theater war und sich am Nachhauseweg vom eben erlebten Theatergenuss verführt fühlte, selbst ein Theaterstück zu schreiben. Verführt wird er gerne, der amerikanische Autor mit europäischer Seele – oder doch Europäer mit amerikanischem Leben? Vieles vermischt sich und ist nur als Klischee zu benennen. Es ist nicht leicht zu filtern was den Menschen und Kunstliebhaber Carl Djerassi beeinflusst hat und auf welche Weise sich diese Wirkung vollzogen hat.

Carl Djerassi, der es nicht mehr gerne sieht, wenn sein Name hauptsächlich mit der Erfindung der Pille in Verbindung gebracht wird, auch wenn sie, wie er selbst sagt, ihm ermöglicht hat, sich der Welt der Kunst zu nähern und sich mit Kunst zu umgeben, hat klar vor Augen, dass – wie Wirtschaft und Wissenschaft – auch Wirtschaft und Kunst sehr nah zusammen gehören. Diese Verbindung ist in den Vereinigten Staaten sehr prononciert: Sponsoren sehen nach zwei Wochen Proben ein paar Szenen an und entscheiden dann, ob sie ihr Geld für diese Produktion ausgeben wollen oder nicht. Ich habe von Sponsoren gehört, die den Ausgang eines Stückes mit bestimmt haben. In Europa ist das *noch* undenkbar, aber

„Schönberg auf dem Parnass", Arnold Schönberg Center, Wien 2008
©Claudia Prieler

die Amerikanisierung Europas schreitet auch auf diesem Gebiet stark voran. Noch sind wir Europäer begünstigt, was diese enorme Abhängigkeit des kulturellen Lebens von der Wirtschaft betrifft und das trifft auch auf das Theater zu. Dies ändert sich jedoch mit der Verschlechterung der wirtschaftlichen Lage. An der Kunst wird, wie allgemein bekannt, zuerst gespart.

Europa nähert sich U.S.-amerikanischen Finanzierungsmodellen an – und damit der Abhängigkeit von der Börsenwirtschaft. Kunst wird zur Aktie – wie schade für die europäische Kultur, denn das Verständnis und die Ausübung kultureller Tätigkeiten kann nur mit Blick auf Wirtschaftlichkeit nicht funktionieren. Kunst kann sich ohne freien künstlerischen Ausdruck nicht entwickeln.

Wir wissen wie abhängig die Wissenschaft von der Finanzwelt ist; wie hoch die Beträge sind, die da fließen. Wir am Theater brauchen auch Geld für unsere Tätigkeit, und wir hätten gern mehr, und mehr Akzeptanz, weil doch für eine Gesellschaft, die sich weiter entwickeln und nicht stagnieren soll, Kultur unerlässlich ist. Elfriede Jelinek hat das bereits in *Die Kontrakte des Kaufmanns* (2009) thematisiert; in dem Stück kommen nun schon „das Geld" und „der Markt" als Protagonisten zu Wort: „Die werden es einmal besser haben." heißt es bei Elfriede Jelinek, „Wir nicht!"

Aber noch ist unsere europäische Arbeitsweise, bedingt durch die Strukturen, in denen wir leben, anders als in den Vereinigten Staaten, wie mir auch bei der

Premiere von Carl Djerassis *Phallacy* in New York City von der Regisseurin und den Produzenten bestätigt wurde. Ebenso in Berkeley oder bei Festivals in Europa mit amerikanischen Gastspielen. Ich habe Amerikaner über Gastspiele aus Europa sagen hören: „Kommt schon wieder so ein Euro Trash zu uns, können die keine Stücke mehr normal aufführen!?" Und man hört die Europäer entgegnen: „Sind die konservativ, die Amerikaner! Zum Broadway gehe ich gar nicht mehr. Dieses fantasielose Spielbein-Standbein bewegen, dieses Sitz- und Stehtheater ohne Tiefe und politischen Anspruch, das hat es bei uns vor 80 Jahren gegeben! Und auch da nur in der Oper!"

Diese Klischees stimmen natürlich nicht (immer). Ich habe etwa in Berkeley ein sehr gut gespieltes Stück zu *Tausendundeine Nacht* gesehen, *The Arabian Nights* (Drehbuch und Aufführung von Mary Zimmermann), das sehr modern und fantasievoll gemacht war. Und ich habe in New York in kleineren Theatern wirklich interessante Abende erlebt. Aber wenn etwas wirklich neu war, kam es aus Schottland oder England oder es war eine Gruppe wie die in New York formierte Wooster Group, die mittlerweile ja auch schon etabliert ist, aber sich immer von der Tradition des amerikanischen Theaters absetzten wollte und sehr innovativ war und immer noch ist.

Wie wirkt sich aber nun ein Leben in Amerika auf diesen aus Europa stammenden Menschen/Wissenschafter aus, der sein Leben zwar mit Kunst verbringt, aber das nur *neben* seiner Karriere als Wissenschafter, welche Priorität hatte und sehr zeitaufwendig war. Seine Zeit verbringt Carl Djerassi heute anders, er will nicht mehr ohne europäische Kultur leben, und fragt sich selbst, warum die Menschen in San Francisco nicht öfter ins Theater gehen bzw. warum es nicht mehr Theater gibt oder warum die Oper nur ein paar Monate im Jahr Opernaufführungen spielt. Er gibt sich selbst die Antwort, die damals, als er noch nicht Theaterstücke geschrieben hat, auch seine eigene war: Arbeit. Die Menschen haben keine Zeit. Das Berufsleben mit seiner Unrast fegt über die Kultur hinweg.

Wenn ein Schriftsteller, der Biografien oder Memoiren oder Theaterstücke schreibt – und darin sind sich, glaube ich, alle einig – immer autobiografisch schreibt, so ist die Andersartigkeit, die Eigenheit, das Unperfekte, Unklare, das, was nicht zusammen passt, das er während seines Lebens mit sich selbst und mit anderen in geografischen und kulturellen Bereichen entwickelt hat – kurzum alles womit der Perfektionist und Wissenschafter Carl Djerassi im Clinch steht – sicher aber als Grundstein für Ausdruckskraft eines Schriftstellers und seine Art der Vermittlung zu nennen. Immer wieder erwähnt Carl Derassi in seinen Interviews oder in seiner Biografie, dass monologisches Schreiben als Konvention für wissenschaftliches Schreiben definiert wurde und Naturwissenschafter daher selten in der Lage seien, sich dialogisch auszudrücken. Aber nur weil Wissensvermittlung in aufgeteilter Rede stattfindet, ist es noch lange keine Theaterszene. Will heißen: dialogisches Schreiben allein ist noch keine Dramatik.

Sicher ist auf jeden Fall, dass Carl Djerassi sehr gut im Monologisieren ist. Es gibt keinen Interviewer, der nicht glücklich darüber ist, Carl Djerassi interviewen zu dürfen. Es genügt eine einzige Frage, das weitere Interview führt dann Carl Djerassi mit sich selbst – wie mir Interviewer immer wieder erzählt haben. Das ist zwar nicht typisch europäisch oder amerikanisch, aber unter Umständen typisch Wissenschaftler.

Sicher ist allerdings auch, dass Carl Djerassi sehr interessante Themen wählt und hervorragende Dialoge schreibt. Trotz dieser Tatsache behaupte ich aber, dass es Naturwissenschaftlern sehr schwer fällt sich dialogisch *emotional* auszudrücken, da für sie auch der Dialog die Aufgabe des Vermittelns von Information hat. Natürlich findet man – auch sehr berühmte – Monologe in Theaterstücken; aber sie vermitteln nie selbst distanziertes Wissen. Diese ist für uns Theaterleute immer nur *vordergründiger* Beweggrund einer Figur. Monologe können politische und gesellschaftspolitische Statements sein, die den Handlungsverlauf eines Stückes komplett verändern; aber sie *beschreiben* in erster Linie – und ich sage bewusst nicht „erklären" – die inneren seelischen Zustände eines Charakters: der innere Monolog.

Was wir Regisseure also verzweifelt suchen – und gerade bei Stücken, die didaktisch, geschrieben sind, und die Stücke von Carl Djerassi sind didaktisch, erklärend, informativ und pädagogisch geschrieben – ist jener Strohhalm, der uns emotional gesteuerte Vorgangs- und Denkweisen der Figuren entdecken und *miterleben* lässt, um die Beweggründe von Figuren aufzudecken und zu erfassen und erfahren, was denn eigentlich wirklich *hinter* der Vermittlung ihres Wissens steht. Das bedeutet im weiteren Sinn: der Figurendramaturgie auf den Grund zu gehen; will heißen: die Psychologie der Figur und den Beweggrund für ihre Handlungen zu finden und manchmal auch zu *er*finden.

Man kann es auf eine einfache „Formel" herunter brechen – wobei Formel nicht wissenschaftlich gemeint ist – „need" und „action" einer Figur im Sinne des „method acting". Sie wurde vom amerikanischen Schauspiel-Guru Lee Strasberg, basierend auf der Arbeit von Konstantin Stanislawskij entwickelt. Das „emotionale Gedächtnis" sollte zu einer größtmöglichen Identifikation des Schauspielers mit seiner Rolle führen, das heute noch in jeder europäischen Schauspielschule gelehrt wird. Durch diese „Formel" wird Information nicht direkt mitgeteilt, sondern sie ist Ergebnis der Bedürfnisse und der daraus resultierenden Handlungen der Charaktere.

Ein Charakter, eine Figur muss am Theater agieren – und das in einer von ihr als adäquat gedachten, bestimmten Weise. Aber ihr „need", ihr Beweggrund, ihr Bedürfnis, und ihre Notwendigkeit zu handeln, sind sehr unterschiedlich davon, *wie* sie tatsächlich agiert. Das ergibt zwei sich diametral kreuzende Spannungsebenen, und gehört zum Erfinden einer Figur. Im weiteren Sinn funktioniert das in ähnlicher Weise beim Erfinden einer Situation.

„Information zu vermitteln" ist nie ein alleiniger Beweggrund am Theater. Die Figur vermittelt Information „um zu" – *um* ihr Verhalten akzeptiert *zu* finden, *um zu* überreden, in jedem Fall aber *um* eine bestehende Lebenssituation in der sie sich befindet *zu* verändern, und zwar indem sie agiert, nicht diskutiert. Und nicht, indem sie ‚um das Agieren herum redet'. Das Reden *muss* ersetzt werden durch *Tun*, und dieses *Tun* mit *Emotion*. Nur dann habe ich als Zuschauer ein unmittelbares Erleben.

Strasberg hat durch seine Arbeit die amerikanische Spielweise so verändert, wie wir sie heute von allen großen amerikanischen Schauspielkünstlern kennen. Wie viele Europäer, Billy Wilder, Bertolt Brecht, Ernst Lubitsch, und alle anderen, die Europa verlassen mussten, hat auch der Europäer Lee Strasberg die amerikanische Schauspielkultur nachhaltig beeinflusst. Interessanter Weise hat die Umsetzung dieser Arbeit viel weniger im amerikanischen Theater stattgefunden als im amerikanischen Film. Im Film allerdings ist sie einzigartig.

In Europa ist Theater ohne Stanislawski und Strasberg – und auch ohne viele weitere Theaterväter wie Jerzy Grotowski, Bertolt Brecht, aber auch Eleanora Duse, Max Reinhardt und Fritz Kortner nicht vorstellbar. Sicher spielt da auch das Bewusstsein eines langen kulturellen Erbes hinein, das wir mittragen. Die europäische Kulturgeschichte, russische, polnische Schriftsteller und Theatermacher, die Comedia dell'arte, spanische, französische, deutsche und österreichische Autor/innen lassen unsere tiefen emotionalen Wurzeln erkennen, die ja auch Carl Djerassi mit seiner europäischen Herkunft erlebt und empfunden hat.

Als Europäer, der in den Vereinigten Staaten mit der Pille Geschichte geschrieben hat, will Carl Djerassi nun sein amerikanisches Leben in seiner ehemaligen europäischen Heimat anerkannt sehen. Er will wahrgenommen werden, mit allem was er geleistet hat – und das ist bekanntlich ja nicht wenig – indem er nun in Europa Geschichten schreibt. Es sind amerikanische Geschichten. Seine Neugier führte ihn eben wieder etwas Neuem in seinem Leben: Wissenschaft im Theater zur Diskussion zu stellen.

Wie man empfindet, vertrieben worden zu sein, aus einem Europa „hinaus geschmissen" worden zu sein – wie Djerassi sagt – das hat er in seinem Buch *Vier Juden auf dem Parnass* (2008) ausdrucksstark zur Diskussion gebracht. Es beschreibt die Kraft sich durchzusetzen und er beweist, dass seine Figuren nicht leicht zu bezwingen waren und sind. Ebenso wie er selbst. Djerassi bezeichnet *Vier Juden auf dem Parnass* als „Doku-Drama" und nicht als Theaterstück. Aber in der Art und Weise wie es geschrieben ist, unterscheidet es sich tatsächlich nicht wesentlich von seinen Theaterstücken. Es ist in jedem Fall sein bestes Dialogstück.

Zu welchem Zeitpunkt in seinem Leben hat er es geschrieben? Und welche Entwicklung hat der Prozess des Schreibens in ihm, dem Autor, ausgelöst? Die Identitätsfindung eines Menschen, der gewaltsam aus seiner Heimat ‚geschmissen' wurde, die Suche nach wahrer Heimat und das Ankommen – wobei die Frage nach ‚wahrer Heimat' ungelöst zu bleiben scheint – sind beständige Themen im Leben eines solchen Menschen. Aber in Amerika hat er sie gesucht und gefunden, sowohl privat als auch beruflich. Dennoch ist er seit einiger Zeit auf dem Weg, sie in Europa wieder zu suchen – ebenso wie seine Protagonisten in seinem Buch.

Er hat das Schicksal des Emigranten erfahren, der in keiner Sprache ohne Akzent spricht (wie er von sich selbst manchmal sagt), und von dem deutschsprechende Theaterleute sagen: „In der englischen Sprache, in der er ja schreibt, sind sein Stil und seine Ausdruckskraft sicher am Punkt" während englische Theaterleute zu seinem Stil und seiner Sprache sagen: „Er muss wohl deutsch bzw. wienerisch denken".

Der persönliche Bezug zu den in seinem Doku-Drama *Vier Juden auf dem Parnass* besprochenen Themen zwingt dazu, die eigene persönliche Frage nach Identität neu zu überdenken. Das Buch, der Text, ist ein emotionaler Schlagabtausch: direkt, emotional, wahr und auch autopsychoanalytisch? Wächst man, wenn man sich besser erkennt, Ängste überwindet, Wahrheiten zulässt, für die man Abstand zu seiner eigenen Entwicklung gebraucht hat? Entwickelt man sich, wenn man sie während des Schreibprozesses beschreiben und auszudrücken lernt? Carl Djerassi bejaht diese Frage.

Außer bei *Vier Juden auf dem Parnass* bin ich mir als Regisseurin fast aller seiner Stücke noch nicht sicher, wo genau sich diese europäischen Wurzeln in seinen Theaterstücken, die er mit dem Überbegriff *science-in-theatre* bezeichnet, befinden. Er ist eindeutig ein amerikanischer Dramatiker, weil er seine wichtigsten Lebenserfahrungen privat und beruflich in Amerika gemacht hat und sich seine Themen auf sein amerikanisches Leben beziehen. Das betrifft selbst *Ego*, das einzige Stück, das nichts mit Wissenschaft zu tun hat.

Vordringlich, und in allen Stücken, geht es ihm um das Konkurrenzverhalten in der wissenschaftlichen Gesellschaft. Sind nun Wissenschaftler interessanter und ihre Konkurrenzgeschichten dramatischer als die anderer Charaktere? Nicht mehr und nicht weniger als die Schicksale von Vätern, Kindern, Priestern, Hausmeistern, Königen oder Frauen. Auf ihren Charakter und ihre Geschichte kommt es an, ihr Dilemma, ihre Psychologie.

Das betrifft auch das bei Djerassi prominente Thema Frauen in der Wissenschaft. Sie haben kaum andere Probleme als Frauen am Theater. Die Probenzeiten am Theater nehmen keine Rücksicht auf die Familie, wir arbeiten bis spät in den

Abend, sind heute hier und morgen dort. Es wäre hilfreich, wenn Männer zu Hause blieben, aber Männer verdienen immer noch mehr als Frauen und diese Tatsache ist im entscheidenden Moment oft der Grund dafür, dass die Dinge so bleiben, wie sie waren.

Carl Djerassis privates Frauenbild mag europäische Kindheitserinnerungen aufkommen lassen, aber insgesamt ist es ein amerikanisches, geprägt von einer Zeit großer gesellschaftlicher Veränderungen, die er wiederum selbst durch seine wissenschaftliche Errungenschaft (die Pille) mitgeprägt hat. Es war eine vehemente, sehr um Freiheiten kämpfende Zeit und eine in Freiheiten lebende und lebendige Zeit.

Die Frauen – Hauptfiguren in seinen Theaterstücken – sind für Djerassi oft die ‚besseren Menschen'. Dabei sind sie aus sehr persönlichen Gesichtspunkten beschrieben und damit auch aus einem rein männlichen, keinesfalls objektiven Blickwinkel betrachtet. Diese Frauen haben – von moralischen Gesichtspunkten aus gesehen, mehr Fehler als die Männer – etwa Melanie – in *Unbefleckt* oder Mme Lavoisier in *Oxygen* oder auch Regina Leitner-Opfermann in *Phallstricke* und Miriam in *Ego*. Sie sind diejenigen, die moralisch unrichtig handeln, sogar wissenschaftlich unrichtig, und ihre Handlungen sind für die Männerwelt – von einem männlichen Standpunkt – nicht nachvollziehbar.

Sie sind bessere Menschen nur unter einem ‚romantischen' Gesichtspunkt, der die Gefühle in den Vordergrund stellt. Wenn man diese Frauen und ihr Verhalten betrachtet, so muss man feststellen, dass sie in der im Stück beschriebenen Zeit nicht reüssieren und auch in unserer heutigen Zeit nicht reüssieren würden. Fehler sind Fehler und führen nur in (amerikanischen) Komödien zu einem glücklichen Ende mit Kompromissen – nicht in der Gegenwart. Sind Frauen nun Hauptfiguren, weil es Djerassis verdeckter Wunsch ist, dass sie die unlogische und irritierende Gefühlswelt nicht vergessen sollen, weil sie doch diejenigen sind, die den Auftrag haben, diesen Teil der Geschichte zwischen Mann und Frau zu erhalten, und ihn auch weiterhin abdecken sollen?

Die Frauen in seinen Stücken sind vielleicht ‚menschlicher' als die Männer im Sinne der Fehler, die sie machen, im Sinne dessen, dass sie nicht vollkommen sind, ambivalent, undurchsichtig, getrieben von unterschiedlichsten Sichtweisen und Aspekten des Lebens, oft von Emotionen gesteuert, eben unwissenschaftlich. Und wenn alles biografisch ist, ist das dann vielleicht als ein versteckter Wunsch des Carl Djerassi zu sehen, nicht nur getrieben sein zu wollen, linear den Workoholic zu leben, sondern seinen Fehlern ins Auge zu sehen, die er nur zu gerne offen begangen *hätte*, die ihm sein Ehrgeiz und seine Konsequenz im beruflichen Sinn aber nicht gegönnt haben? Hätte er gerne diese weibliche Seite mehr gelebt? Ist das seine sich erinnernde ‚europäische' Seele?

Diese Auseinandersetzung zwischen dem amerikanischen, ehrgeizigen Wissenschaftler, der Wissen und Errungenschaft vermitteln will, und dem sich sehnenden Europäer ist in seinem Schreiben als wesentlicher Punkt zu erkennen. Es ist sehr wichtig für die Dramatik seiner Stücke, die das Leben der von Ehrgeiz besessenen („poisoned", vergiftet, wie Djerassi selbst gesagt hat) wissenschaftlich arbeitenden Menschen beleuchten. Wobei sich die Frage stellt, ob er, als Teil dieser Welt, ebenso mit sich selbst ins Gericht geht?

Wie funktioniert also Wissenschaft am Theater, und wann? Nehmen wir als Beispiel das Stück *Unbefleckt* und zwar jene Szene, in der das erste Experiment in der Geschichte der künstlichen Befruchtung mittels ICSI gezeigt wird. Warum funktioniert diese Szene? Weil sie einen starken persönlichen emotionalen Hintergrund der Hauptfigur Melanie hat, die von persönlichen Motiven regelrecht in diesen wissenschaftlichen Selbstversuch hinein gepuscht wird; also: „need" und „action". Oder eine weitere Szene, in der Menachem, der immer geglaubt hat, unfruchtbar zu sein, plötzlich erkennt, dass er Vater werden kann bzw. schon geworden ist. Es ist also wieder ein persönlicher Beweggrund der Figur. Und da Carl Djerassi, wie bereits gesagt, wirklich gute Dialoge schreibt, ist diese Szene auch eine wirklich unter die Haut gehende – und komödiantisch ist sie noch dazu.

Die Wissenschaft ‚funktioniert' auch in dem Stück *Oxygen*, das Carl Djeassi und Roald Hoffmann gemeinsam geschrieben haben, solange es um den Kampf der drei Hauptfiguren Lavoisier, Priestley und Scheele geht; um die Frage, wer den Sauerstoff zuerst entdeckt bzw. verstanden hat. Doch in dem Stück gibt es eine Szene – und was tut man mit so einer Szene auf der Bühne? – die mindestens 20 Minuten lang ist und nichts anderes zeigt als ein Experiment, das noch dazu immer wieder wiederholt wird, wenn auch mit leichten Abänderungen.

Carl Djerassi hat große Freude daran, Wissenschaft ins Theater zu ‚schmuggeln'. Aber das ist keineswegs Verführung des Publikums durch Wissenschaft. Vielmehr ist es ganz offensichtlich didaktisch, vielleicht für manche auch lehrreich, für einige – unter den Schauspielern und im Publikum – ist die Wirkung allerdings auch gnadenlos.

Wie setzt man also eine solche Szene auf der Bühne um? Die theatrale Realität muss ja erhalten bleiben, und die Dramatik darf nicht aufhören, nur weil ein wissenschaftliches Experiment gezeigt wird, das per se am Theater nicht so dramatisch ist wie ohne Zweifel für einen Wissenschaftler im Labor, selbst wenn es die Entdeckung des Sauerstoffs betrifft. Mein Versuch theatralischer Umsetzung dieser Endlos-Experimentier-Szene bediente sich des Puppentheaters und Showeinlagen.

Doch selbst diese sich theatralischer Mittel bedienende Umsetzung funktionierte nur, weil ich den Verlauf des Stückes als Regisseurin dahingehend verändert habe, dass die Hauptfiguren des Stückes, welche Mitglieder eines Komitees

Oxygen, Mainfrankentheater Würzburg, 2001
©Regine Körner

zur Vergabe eines Retro-Nobelpreis für die erste bahnbrechende Entdeckung in der Naturwissenschaft sind – ihre Kandidaten aus dem Jahr 1777, Scheele, Priestley und Lavoisier selbst spielen, und zwar so wie sie sich mittels ihrer Fantasie *vorstellen*, dass Scheele, Priestley und Lavoisier damals agiert haben *könnten*. Der Fantasie ist also auf dem Boden des dramaturgischen Stücks und der freien Erfindung der charakterlichen Entwicklung der Figuren freier Lauf gelassen. Das geht natürlich mit Veränderungen am ursprünglichen Text einher, da der Text Situation und Charaktere unterstützen muss – und nicht erklären.

Fantasie muss am Theater stattfinden; das ist einer der großen Unterschiede zum Pädagogischen, Didaktischen, zur haarscharfen wissenschaftlichen Abbildung. Und der Zuschauer muss emotional am Leben der Figuren des Theaterstückes direkten Anteil nehmen und teilhaben können, nicht nur am Wort.

In Tom Stoppards *Arkadien* (1993) findet eine Unterrichtsstunde statt und nicht das Publikum, sondern die Schülerin wird belehrt, die selbst ein altkluges Kind ist und den Lehrer emotional total fordert. So ein Kind hätte doch jeder gern: hübsch, liebenswert, und unglaublich klug. Und die Szenen sind dazu auch noch witzig. Es geht im Grunde darum, wer wen betrogen hat und wer diesen grässlichen Schriftsteller denn nun umgebracht und was Lord Byron mit all dem zu tun hat. Also ein Boulevardkrimi mit wissenschaftlichen Zutaten.

Oder das Stück *Kopenhagen* (1998) von Michael Frayn. Er wechselt ganz bewusst zwischen den Zeiten, während er die historische Aufarbeitung der Beziehung zwischen Heisenberg und Bohr zum Thema macht, aber mit ungelösten Antworten, die den Zuschauer in Spannung halten und mit Möglichkeiten füttern, die die Phantasie anregen. Ist Heisenberg nun ein Verräter oder nicht? Was wollte er von Bohr? Mit der Atombombe und Atomgegnern hat man doch einen unmittelbaren Bezug zum eigenen Leben und die Heisenberg'sche Unschärfe Relation gehört inzwischen zur Allgemeinbildung. Allerdings habe ich bei einer Aufführung eines Tourneetheaters auch Leute protestierend aus dem Theater laufen sehen: „Was glaubt der eigentlich, dass ich noch einmal die Schulbank drücke und eine Lehrveranstaltung über Physik bekommen will? Frechheit!"

Interessant ist, dass es Theatermacher immer toll finden, wenn Leute lautstark das Theater verlassen, weil es die schläfrige Masse aufrüttelt und zumindest eine Reaktion auf das ist, was man gesehen hat; Autoren empfinden das aber immer als einen persönlichen Angriff auf ihre Person. Ich habe diese Reaktion jedenfalls als eine bemerkenswerte Aktion wahrgenommen. Tatsächlich ist ein Stück wie *Kopenhagen* in London ein Erfolg gewesen, aber in keinem Staatstheater oder einem anderen großen deutschsprachigen Theater aufgeführt worden. Die Dramaturgen und Theaterleiter hatten Angst, dass das Thema zu trocken ist, zu didaktisch, dass es niemanden interessieren würde, zu lehrreich sei, nur Wissenschaft vermittelnd, einer Vorlesung gleich.

In Deutschland hat es eine Tournee gegeben und sonst nur Aufführungen in kleinen Theatern, die Bearbeitungen und starke Kürzungen vorgenommen haben. Sie haben das Stück umgeschrieben, um das Stück theatralischer und kurzweiliger zu machen und in jedem Fall der Tradition des Konversationstheaters etwas entgegen zu setzen bzw. etwas hinzu zu fügen. Ob das eventuell auch damit zu tun haben könnte, dass die deutsche Sprache an sich schon komplizierter ist und durch komplizierte Satzbauten einen eher erklärenden Duktus hat? Und ob Englisch die bessere Sprache für das Genre ist? – diese Frage möchte ich nur als Möglichkeit in den Pot werfen.

Ein Hauptgrund aber, warum diese Stücke mit wissenschaftlichem Inhalt bzw. Thema so selten gespielt werden, ist der Folgende: Theater ist *Illusion*. Bewältigung von Illusion. Auf einer Theaterbühne wird eine Geschichte erzählt, die erfunden und fiktiv ist, und die ich als Zuschauer im Moment als wahr nachvollziehen kann. Eine Geschichte – selbst von Personen, die real gelebt haben, nehme ich im Moment als eine andere Wahrheit wahr und empfinde sie im Moment. Publikum und Akteure glauben in demselben Moment an dasselbe. Ich überzeuge das Publikum, dass das was auf der Bühne passiert, *wirklich* ist, aber es ist Illusion. Erfundenes wird im Theater zur ‚Wahrheit'. Theater ist nicht Erklärung der Realität – oder Beweis der Realität. Es geht um keine exakte Abbildung.

Die Wissenschaft gibt mir diese Freiheit zum willkürlichen Erfinden nicht.

Wissenschaft vermittelt eine Tatsache, ein Faktum, das auf der Bühne dargestellt wird. Sie beweist. Ist exakt. Zuschauer wollen aber mitgenommen, verführt, emotional überredet werden, auch Rückschlüsse auf Situationen in ihrem eigenen Leben machen können. Dann lernen sie gerne. Didaktisches, pädagogisches Vermitteln von Wissen tut das nicht – zumindest bei einem großen Prozentsatz von Zuschauern. Ich muss eine Situation dramatisieren, die ich beschreiben will. Und dramatisieren heißt nun einmal auch: etwas lebhafter, aufregender darstellen als sie in der Wirklichkeit ist. Dramatisch muss es sein! Was sind Synonyme für „dramatisch"? Dramatisch ist nicht nur Neuartiges und Unerwartetes. Spannend soll es sein, und das bedeutet ausdrucksvoll, fesselnd, explosiv, aufrüttelnd, packend, lebendig, mitreißend, wild, derb, kritisch sowie bemerkenswert und oft auch nervenaufreibend und ungenießbar.

Theaterstücke mit dem Thema Wissenschaft werden oft dramatisch bezeichnet in der Bedeutung von interessant, informativ, relevant, sehens- und hörenswert, geistreich, lehrreich, aufschlussreich, erzieherisch, wissenswert, aufklärend, wichtig und erhellend – aber selten aufrüttelnd oder emotional extrem fordernd. Ich behaupte, dass all diese Stücke nur szenenweise funktionieren, und nicht als gesamte Stücke. Werktreue ist ja auch in diesem Fall kein Garant für gutes Theater!

„Es hat mich nicht berührt" – ist oft eines der Kriterien mit denen man eine Aufführung negativ bewertet. Theater hat zusätzlich mit Freizeitgestaltung zu tun, mit Ausspannen, mit lachen können. Ist es lustig? – ist bei Kartenbestellungen oft die erste Frage.

Nur Unterhaltung? Das will Djerassi eben auch nicht: „I want the people to be educated not only to be amused in the theatre."

Es ist für einen Wissenschaftler sicher unglaublich aufregend, sich im Prozess des Entdeckens zu empfinden, in dem Moment, in dem sich ihm seine Entdeckung eröffnet und erhellt; sich ihm das Unerwartete, Neue als möglich offenbart. So ist auch am Theater *die Unmittelbarkeit, die Unmittelbarkeit des im Moment stattfindenden theatralen Erlebens* von größter Wichtigkeit. Wenn es nur erzählt wird, ist es schon vorbei.

So absurd das klingt: Gut sind Stücke über Wissenschaft immer dann, wenn das Publikum nicht verstehen muss – kann, aber nicht muss! – wie ein wissenschaftliches Experiment oder eine These funktionieren. Sie sind gut, weil die Figur im Stück mit dem Experiment etwas anderes zu vermitteln versucht als das Experiment selbst – nämlich den Zustand seines Charakters, seiner Figur, die Situation in der sie sich befindet, und somit das Experiment nur als Beweggrund dafür verstanden werden kann, dass diese Figur versucht, ihren persönlichen Zustand zu ändern; und die Stücke sind gut, wenn ich psychologisch interessiert und gleichzeitig emotional „erwischt" werde.

In dem 2009 in Wien uraufgeführten Stück *verrechnet* von Carl Djerassi und Isabella Gregor, will die Figur der jungen aufstrebenden Schauspielerin Polly dem Leiter des Theaters damit imponieren, dass sie ihm anhand eines einfachen bildhaften Beispiels, mittels eines Apfels, die Differentialrechnung erklärt. Ihr Beweggrund ist allein der Beweis, dass sie ehrgeizig und gut ist, und endlich in das Theaterensemble aufgenommen werden muss.

Obwohl Wissenschaft weltweit unsere Gesellschaft beeinflusst und bestimmt – scheinen Theaterleute eine Art Wissenschaftsphobie zu haben. Seneca sagt: *Saepe bona materia cessat sine artifice* – oft bleibt ein guter Stoff ungenutzt, weil der Künstler fehlt. Das Theater sagt: Saepe bonus artifex cessat sine materia. Oft bleibt ein guter Künstler ungenutzt, weil die Materie fehlt. *Let's start to communicate!*

verrechnet!, Stadttheater Walfischgasse, Wien 2009
©Moritz Schell

Carl Djerassi's Science-in-Theatre Plays: The Theatrical Realization

Andy Jordan

In the late spring of 1998, I was a Senior Radio Drama Producer/Director with the BBC in London. I took a call from a playwright friend of mine who asked if I wanted to meet one of the world's most renowned scientists who, perhaps surprisingly, had written a play. The play was going to be presented at the Edinburgh Fringe Festival in August of that year, and the young, inexperienced director was keen to find a theatre producer with experience at the Edinburgh Festival to act as a consultant on the production. My friend wondered if I would be interested in being that consultant. I was. This was how I first made Carl Djerassi's acquaintance, which led to my directing a new version of his play, *An Immaculate Misconception*, for BBC World Service radio drama in 1999.

Having worked as dramaturg, director and producer on the development and production of all of Carl Djerassi's theatre plays in the UK, I have a unique perspective on them. I was also responsible for the directing and producing of radio drama versions of two of the plays on BBC World Service. With direct reference to a number of Djerassi's plays, and to my productions of them, this article will cast light on the artistic collaboration between a professional theatre practitioner (with no scientific training whatsoever) and a scientific genius-turned-playwright. Along the way, I will also examine some of the chief preoccupations of, and recurring themes in, Djerassi's plays. I mention this because what the plays were about, and how they worked as drama, was the bedrock of our work together. Indeed, one of the privileges and excitements of working on new plays (as opposed to revivals of plays which have already been produced) is that one is often the first person to deal with the play's subject matter and its themes.

Almost all of my professional work in the theatre and in radio drama has involved me in directing and producing new plays, usually by new writers. A 'new play' is when the play is given its first-ever production, what is called the world premiere. It may be useful in this context to reflect on the fact that Shakespeare was once a new writer, as were Harold Pinter, Tom Stoppard, Arthur Miller,

Samuel Beckett and Alan Ayckbourn (the most popular living playwright in the world today). Somebody had to nurture and encourage these once new writers who were showing promise and potential as playwrights. That 'somebody' has normally included directors and producers. Somebody has to show belief in a new writer and take a chance on them by giving them a break by putting their play on. In the professional context, this act of risk on the part of directors and producers is very meaningful: after all, when a writer is 'new', the public has never heard of either the writer or the play, which makes promoting new plays a hugely difficult task. In truth, it is a brave, and committed, producer who takes that risk, and without such producers and directors the world would never have known about Shakespeare, Pinter, Stoppard, Miller or Ayckbourn. Or, for that matter, in the context of playwriting, Carl Djerassi, for Djerassi, too, was once a new, unknown playwright.

When I was first introduced to Carl Djerassi I was especially intrigued by the fact that such a remarkable scientist wanted to become a playwright. After all, don't many people believe that science and the arts are in some way antithetical, and in permanent opposition? As it happened, I had always been interested in plays about science, as I was engaged by their potential political and social impacts. I had commissioned, directed and produced my first theatre play about a science subject as far back as 1981, Paul Unwin's *Theory for the Attention of Mr Einstein*, about which a critic said: "... a glittering edifice of ideas ... the acting is excellent: a Nobel Prize for approaching the unapproachable, and getting impressively near." (Hughes 37) I also worked on various science-related projects in radio drama, including commissioning and directing Paul Thain's *The Paradise Machine*, about developments in virtual reality, which went on to win the 1995 European Broadcasting Union Award.

In discussing my involvement in the directing and staging of Carl Djerasssi's plays, the fact that I was an experienced exponent of working with new writers on new plays is relevant. I believe this was one of the factors that interested Djerassi in me as a potential director of his plays. That I had had some success working on the development, directing *and* producing of new plays was, I am sure, important to Djerassi, as indeed it should have been. Given this, it might be useful for me to explain some of what is involved in this process of nurturing and developing new writing, and new writers, in the theatre.

My entire working life has been spent staging and producing original plays, with both new and established writers, presenting the premiere productions of them. Many of the writers I have 'discovered' (by which I mean that I was the first director/producer to present one of their plays), and whose work I have helped find an audience for, have gone on to become internationally successful. Directors of new plays inevitably develop close working relationships with writers, especially if the relationship extends over a number of productions as mine and Carl Djerassi's has. These relationships will inevitably be different between each writer and director, but the nurturing of script and writer is almost always a key part of the journey of getting a new play from page to stage. This type of work can take numerous forms, from close quarters analysis (using script reports, sometimes prepared by playreaders or script editors, discussion, re-writing) to workshopping (having a director and actors work on a new play in workshop mode, testing the script in action in the rehearsal room) to staged and rehearsed readings of a play before an audience (followed by feedback from the audience for the writer, director and actors).

My working relationship with Djerassi has developed over the years, becoming increasingly more practiced and efficient. When we began working together, Carl Djerassi was new to playwriting, and was honest about it being a skill he needed to improve on, something he has avowedly long since managed to do. I would say the relationship I have with Djerassi is probably closer than I have ever had with any other writer. There are other playwrights with whom I have had (or have) very close working relationships. But because I was director, producer *and* dramaturg, my relationship with Djerassi was bound to be exceptionally close. With many other writers the boundaries (rightly so) are clearer between the roles of the director, writer, producer, and the dramaturg. But having to combine and merge the usually separate artistic disciplines of directing, producing and dramaturgy makes my work with Carl Djerassi more interesting and challenging. Perhaps for Djerassi also, as he seems to suggest in an interview in *The Dramatist* magazine in November, 2001: " ... when you finish a play to your satisfaction, the process has only started. The director, dramaturgs, and actors participate. Suddenly it becomes a process of collaboration – and, to that extent, a replication of the scientific process." (Devins 28) An intriguing thought, to compare the two forms of 'creative' process, and one I will return to later.

I have emphasized that what the plays *were about* and how they worked as drama, was the bedrock of our work together. Examining these factors is wrapped up in the word 'dramaturgy' (which also, of course, has various other meanings and dimensions to it). Dramaturgical imperatives were therefore absolutely key to the task at hand. This process always begins with a close analysis of the play, and detailed discussions between writer and director in the hope that some useful insights will arise which can be taken into the re-writing process (after all, as the adage rightly has it, 'writing is all about re-writing'). In my work on Djerassi's plays there was an equality of focus between the dramaturgical development and imperatives connected with staging the play. In regard to dramaturgy, important issues can include: the sharpening of the narrative and thematic arc of the dramatic action; the honing of a dynamic dramatic structure; the meaningful, expressive realisation of character relationships, and the developing of individual character journeys; as well as considerations around theme and message. Whilst I am contemplating dramaturgy, it is also important that I mention an esteemed colleague of mine, Darren Tunstall, a wonderfully talented actor, playwright and teacher who has worked as chief dramaturg with myself on four of Djerassi's scripts, and has become an important part of the team of professionals in the UK who have brought Djerassi's plays to stage life.

When Carl Djerassi and I first began working together, he had spent his life in science and education. Latterly he had been writing a sequence of *science-in-fiction* novels, but he then had become committed to the idea of writing plays. His interest in writing dialogue, and using the dialogic approach as a way of imparting scientific ideas, was therefore already underway. The fact that the first few plays we worked on together (for theatre and radio) were plays about science and scientists, a subject about which he knew much, made for an interesting admix as I knew about directing, producing, and the writing and development of new plays, which was the territory Djerassi had decided to launch himself into. He was comfortable with the knowledge that he needed support from professional theatre practitioners such as myself as he was determined to become a good playwright. Carl Djerassi is not, after all, the kind of man who settles for anything less than excellence and success.

The first two plays of Djerassi's that I directed and produced were pedagogical and didactic (intentionally so, it must be noted), sometimes rather too factually-based, with an argument built around a particular set of incidents and characters (contemporary and historical), which Djerassi hoped would get the audience thinking about the issues. A large component of my role as director-dramaturg was to find a way of making the pedagogy 'theatrical.' In my opinion it is perfectly possible for pedagogy or the didactic to function in drama. If this were not the case, playwrights such as Bertolt Brecht, John Arden, Edward Bond and David Hare would never have had the impact they have had on world theatre. The trick is to ensure that the play's genre and style is clear and consistently adhered to (and, of course, that the acting, directing and production work is of the highest quality possible).

The way the 'creatives' on the productions (such director, producer, set and costume designer, lighting designer, composer, sound designer) approached the 'issue' of the pedagogy, in theatrical terms, was through the realisation of visual images and metaphors (stage pictures), fitting language to period and style, the scenography (the stage design), proxemics (the 'blocking,' stage geography and movement), music, video and film, sound and lighting, and the overall stylistic approach of the production (such as the creation of 'the world of the play,' the quality of acting, the expression of the character's needs in relationship with the other characters). It was our job to make these elements come together to best realise the thematic, intellectual and emotional content of the play in performance for the audience. Sometimes we needed to make the scientific ideas behind the writing even more organic to, and integrated with, the play's dramatic action and themes; occasionally, we needed to further 'explain' a difficult scientific concept or wanted to enliven information that was perhaps a little too academic or factual. It can be argued, as Kristin Shepherd-Barr does in her book *Science on Stage: From Doctor Faustus to Copenhagen*, that "it is not the quantity of science in a science play that matters, but the quality of its integration: the way in which it figures both thematically and theatrically" (Shepherd-Barr 19), which I think is correct and a requirement we were certainly aware of and actively explored in all the plays. In parallel with this work we would discuss the project from a producing perspective, which theatre to present the play in, which actors might we cast, what the marketing and promotional campaign could look like. Finally, whilst all this work was happening, the dramaturgical input into the development of the plays also continued. As already indicated, it has always been part of my job when working closely with Djerassi as director-dramaturg on his plays to find ways to translate sometimes abstruse scientific facts and notions into engaging dramatic action, devising methods to turn exciting events in the history and philosophy of science into gripping, illuminating and entertaining theatre.

During the past twenty-three years Djerassi has turned from active involvement in scientific invention to active involvement in the world of fiction and the theatre and, in spite of his advanced age, he remains one of the most active people I have ever met. Since boldly going where not many other scientists have ever gone before, he has written a series of what he calls *science-in-fiction* novels and *science-in-theatre* plays. In so doing he has been at the heart of promulgating a fascinating new body of theatre plays, *science-in-theatre*. The term, I suspect, needs some explanation. What is it, exactly? And why does Djerassi write it?

Wanting to engage with a wider, non-scientific and non-academic audience was always one of Carl Djerassi's ambitions for his plays (and my ambition, also, as the producer of the plays). In this regard, he talks about himself as an *agent provocateur*. Djerassi is determined to disseminate what he considers to be vitally important ideas, which is why the plays seek to be provocative and entertaining, wanting to inspire audiences to engage with the issues. He believes there is a widening gulf between the sciences and the other cultural worlds of the humanities and social sciences and that scientists themselves spend preciously little time in attempting to communicate with these other cultures. Djerassi decided to do something about illuminating the scientist's culture for a broader audience, and to do it through a tetralogy of novels in a genre he called "science-in-fiction" – not to be confused with *science fiction*. He said:

And so I call the literary genre in which I work 'science in fiction.' It was important to me to differentiate what I do as clearly as possible from science fiction. For me, the most important difference is that in science-in-fiction all the science or idiosyncratic behaviour of scientists described is plausible. None of these restrictions apply to science fiction. . . . But if one actually wants to use fiction to smuggle scientific facts into the consciousness of a scientifically illiterate public – and I do think that such smuggling is intellectually and societally beneficial – then it is crucial that the facts behind that science be described accurately. Otherwise, how will the scientifically uniformed reader distinguish between what science is presented for entertainment and what is informative? (Djerassi 2001, 165)

Djerassi has been a regular theatregoer all his life. In writing his *science-in-fiction* novels – they are all still in print – he found he enjoyed writing dialogue, partly because as a scientist he was not allowed to use the form. He also recognised that dialogue is a far more accessible and entertaining way to impart ideas. After he started to write his science novels he noticed how relatively few *science-in-theatre* plays there were, which encouraged him "to embark on a trilogy of "*science-in-theatre* plays, where science and scientists are central and the facts impeccably correct." (Djerassi, unpublished) In seeking to define *science-in-theatre* Djerassi says, "By this label I refer to plays in which science or scientists do not just fulfill a metaphoric function . . . . In my plays, what I call the 'tribal practices' of scientists constitute the central focus of the drama, as, for instance, in Michael Frayn's *Copenhagen*." (Djerassi, unpublished)

Djerassi's descriptive phrase – *science-in-theatre* – as a way of defining those of his plays which deal with science and scientific issues, and his frequently made assertion that it is a new genre (and, one presumes, equivalent to comedy, tragedy, musicals, or one person plays), has caused some debate in academic circles. To some critics it might be more accurate, or perhaps more convenient, to call this group of plays a sub-genre, as Eva-Sabine Zehelein does in her indispensible book *Science: Dramatic: Science Plays in America and Great Britain, 1990–2007* (Zehelein 103), or to refer to them as a new *art form*. In considering matters of categorisation, Zehelein comes down on the side of "science plays" as an "umbrella term," and as "science plays," as does Kirsten Shepherd-Barr in her equally indispensible book, *Science on Stage*. (Shepherd-Barr 1) Djerassi has developed a highly pertinent distinction between what might loosely be called "science plays" and his definition of plays "where science and scientists are central and the facts impeccably correct," plays which he also describes as "*pure* science in theatre." (Djerassi 2001, 245) Beyond any argument which may persist around the best and most accurate way to classify plays that deal with science, Djerassi urgently turns the focus onto one of the key thematic purposes behind the writing of his science plays:

Science in my opinion has no ethical dimension. Ethics start with individual scientists. Not only is science neutral, you don't even know what you are going to discover. Yet the moment a scientist starts getting some answers, ethical questions enter. The story of the Contraceptive Pill, the first medical treatment for healthy people, shows that the chain of responsibility is a long one. Basic science is one thing, technology another, even if the division between them is not always clear. At what stage along the line from theoretical idea, to experimental research, to clinical application, does the moral buck stop? (Tyler 14)

This profoundly important question is at the heart of most of Djerassi's plays. Carl Djerassi and I frequently discussed how best to use theatre and theatre plays as a platform for disseminating discussion about scientific issues, and whether or not it was doomed to failure or success. In a lecture about *science-in-theatre* at the Royal Institution in London in 2000, Djerassi made a very revealing statement when he said he had come to the Royal Institution "not to talk about bringing theatre to science, but rather science to the theatre; to answer the question, to answer the question, is 'science-in-theatre' a viable genre, or is the phrase itself a contradiction in terms?" (Djerassi 2001, 249) As opposed to Carl's perspective, I saw *my* job as bringing theatre to science.

Another question Djerassi was pre-occupied with was whether *science-in-theatre*, as a new form of dramatic synthesis, can fulfil an effective pedagogical function onstage, or whether pedagogy and drama are antithetical? We have tried to explore and answer this question in each of the plays and their productions. For many people I suspect the jury is still out on this question. However, it was clear to me once I started work on Djerassi's plays that this issue was hugely significant, for it was at the heart of both the plays themselves *and* of who Djerassi was as a writer. There was no ignoring it; we simply had to get the balance right in each play between pedagogy and didacticism and the 'other things' – the story, the characters and their relationships, the dramatic conflicts, or the way the plays were staged. Interestingly, Djerassi's contention – as one way of starting to answer his question – was that science "is inherently dramatic – at least in the opinion of scientists because it deals with the new and unexpected. But does it follow that scientists are dramatic personae, or that science can become the stuff of drama?" (programme note by Carl Djerassi, performance programme for *Oxygen*, Riverside Studios, London, 2001). In thinking about how to treat with the didacticism in the plays, Djerassi was clear that he needed to accentuate other qualities in the writing: "Instead of starting with the aggressive preamble, 'let me tell you about my science,' I prefer to start with the more innocent 'let me tell you a story' and then incorporate realistic science and true-to-life scientists into the plot." (Djerassi, unpublished)

This emphasis on telling engaging stories is vitally important, and was at the heart of much of the dramaturgical work we focussed on – compelling storytelling is one obvious way in which plays that have a didactic and pedagogical purpose can be made more palatable to a non-specialist audience. Djerassi said: "What is wrong with learning something while being entertained? In other words, why not use drama to smuggle (with a substantial dose of theatricality) important information generally not available on the stage into the minds of a general public?" (Djerassi 2007, 101) Nonetheless, Djerassi conceded that audiences did not expressly come to the theatre to be educated but to be entertained. This is not to deny, of course, that in *most of drama* there is an implicit function which is fundamentally concerned with investigating and illuminating the human condition, and it is through this process we learn about ourselves as human beings. The desire and need to tell ourselves stories, and to share in this process, is at the root of much of human behaviour.

Djerassi's plays which I directed were *An Immaculate Misconception* (for BBC World Service radio drama, 1999, but first performed at the 1998 Edinburgh Fringe Festival and then, in my own production, at the Bridewell Theatre, London, in 2002*); Oxygen* (co-written with Roald Hoffmann, performed at the Riverside Studios, London, November, 2001, and then in a new version for BBC World Service radio drama, December 2001); *Ego* (Edinburgh Fringe Festival, 2003); *Three on a Couch* (inspired by Djerassi's novel, *Marx, Deceased*, performed at the King's Head Theatre, London, 2003); *Calculus* (New End Theatre, London, 2004); *Phallacy* (New End Theatre, and King's Head Theatre, London, 2005); *Taboos* (New End Theatre, London, 2006).

Since 2006, we have worked together on an entirely new version of *Taboos* entitled *Keeping it in the Family* (which is, as yet, unproduced), and Carl's latest plays, *Foreplay* and *Insufficiency* (for production in London in 2012). The plays have had great international success, having been translated into nineteen languages and performed and published around the world (all the plays have been published in German, and only *Ego* remains unpublished in English). Uniquely, *Foreplay* has already been published in three countries before we premiere the play in the theatre (which will be with an entirely different version of the text). I conceived and organised the entirety of each part of these projects. This included researching the subject(s), offering dramaturgical advice about the plays, editing the script(s), conceiving, rehearsing and directing the creative production, negotiations with relevant organisations (such as theatres, backers, press, media), casting actors and creative teams, conceiving and planning the publicity and marketing campaign, writing publicity materials and copy, and collaborating with the author on how best to present the artistic, intellectual, scientific and pedagogic ideas in the production.

An Immaculate Misconception. BBC World Service Radio Drama, 2000, and Bridewell Theatre, London, 2002.

37 year old research scientist Dr Melanie Laidlaw has problems. Her biological clock is ticking and she wants a child...

As the start of an on-going artistic collaboration and research theme into the presentation of science in theatre and radio performance, I directed and produced, at different times and with different casts, a theatre production, a BBC World Service radio drama production and a rehearsed reading of Djerassi's extraordinarily prescient drama, *An Immaculate Misconception*. The first play in what became a trilogy of *science-in-theatre* plays, *An Immaculate Misconception* is set in the world of scientific experimentation in Assisted Reproductive Technology (ART), which includes IVF procedures such as ICSI (intra-cyto-plasmic sperm injection), which the play explored in some detail, a subject Djerassi returned to in his seventh play, *Taboos*, which looked at the impact of such reproductive technology on society, developments made possible by scientists. The play takes place in the explosive arena of reproductive technology. Who better to write it than the man who is said to have ushered in the sexual revolution of the 1960's? His work was so revolutionary, and his achievements so impressive, that he was the only *living* person to be included in *The Sunday Times* list of *The Top Thirty Men of the Millennium*, a list which also included Dante, Chaucer, Da Vinci, Shakespeare, Marx, Napoleon, Darwin, Freud, Einstein and Lenin (and in which Isaac Newton, incidentally, topped the list).

Thirty-seven-year old research scientist Dr. Melanie Laidlaw has problems. Her biological clock is ticking and she wants a child. The man she loves is married, but not to her. As a high-flying scientist, she needs a womb to try her new cure for male infertility. The solution to all her problems seems simple – test the invention on herself. Things become complicated when the man whose sperm she secretly uses discovers he is to become a father. But before Melanie can claim credit for either her revolutionary new technology or motherhood, she must first overcome the professional designs of her envious rival, research scientist Dr. Felix Frankenthaler, who purposely muddles the experiment. Personal and scientific issues soon begin to interact and collide and Melanie is forced to confront the biological father of her child, who discovers his paternity in the strangest of circumstances. Questions about parenthood continue to raise the emotional temperature around Melanie's controversial pregnancy (whose child is it, anyway?), reaching their climax with the birth of a baby boy. The play is also the story of twin births, the first being the birth of a boy called Adam and the second the birth of a real-life process known as ICSI.

Dr. Laidlaw's invention is not science fiction, it is science fact; intra-cytoplasmic sperm injection is a form of test tube fertilisation where a single sperm is injected directly into an egg, a process which, since 1993, has already helped create over 250,000 babies. The play is therefore about the relationship between scientific invention and the facts of life. Once there were birds and bees, now there is a microscope, one sperm and an egg. No women, no man, not even a bed. Controversial news stories appear every week as a result of the mechanisation of reproduction. For instance, in the UK, Diana Blood stored sperm from her dead husband and then used ICSI to give birth to her and her late husband's baby. ICSI has now become the most powerful tool for the treatment of male infertility. As Chris Green said in the performance programme for our London theatre production of the play, the notion of a perfect child at an ideal time has been a human fixation emotionally, medically and even politically for a long time. How often has the fate of nations hung upon a vital birth where the lack of birth or the birth of the 'wrong' sex has had major political consequences? Now modern science offers such miracles to all, or at least all those who can afford it. For prices between £2,000 and £10,000, the infertile can be made fertile, certain genetic diseases avoided, the sex of a child selected and, as the recent case of the Blood family illustrates, even being dead is no longer a barrier these days to being a father. The era of so-called designer babies and children as life-style accessories has, as Carl Djerassi suggests, finally arrived.

The increasing involvement of science and technology in this most fundamental and intimate aspect of human activity creates difficult moral issues. Questioning whether inventors are also responsible for the ethical consequences of their discoveries is one of the major contentious issues for today's society. The play is therefore a theatrical and intellectual exploration of ethical issues surrounding ART. With this particular play and production, we wanted to provoke discussion of the ethics of cutting-edge reproductive technology, and to inform audiences as to what ART, IVF, ICSI, and genetic manipulation could mean for the future of us all, because there is no more topical or significant issue than this one. Genetics, cloning and assisted reproduction have created a brave new world where sex is no longer a vital part of reproduction. The ethical issues surrounding ART and explored in the play – genetic engineering, eugenics, designer babies – profoundly affect us all. Scientists speak in the tribal language of science, a feature of science that is rarely dramatised, but in this case accurately portrayed by the playwright who is also one of the world's most distinguished scientists. The play sought to bring the language and customs of the scientific world into accessible and exciting dramatic form. It shows us what happens when technology which is powerful beyond the means of controlling ('scientists playing God'?) is put into human hands. Alongside this, it is also a witty and deeply thoughtful piece about the meaning of parenthood (do you have to be the biological parent of a child in order to be a good parent?) and, as a sub-theme (but no less important one), this play (and some of the others) also examines the struggle of women trying to break through the glass ceiling of the male dominated world of science.

The production of the play – the second one I did with Djerassi (in 2002) – allowed me an opportunity to engage in and fully explore specific directorial interests of mine, the use of multimedia, music, and a bold visual creation of the world of the play. None of these things could be achieved, of course, without the contribution of gifted designers (set, costume, video, lighting, sound) and composers. Working with such artists – and of the highest calibre – is one of the joys of being a director-producer. On this production, the set and costumes were imaginatively designed by the remarkable Joanna Parker, the video memorably realised by Nic Sandiland and the haunting music was composed by Iain Dunnett. I also recall this production as being something of a watershed as analogue technology was being phased-out in favour of digital; all the video and sound cues were played off digital platforms.

What made the play/production special beyond even this factor was both its subject *and* its theme – after all, plays about politics, history, religion and philosophy have been around as long as drama itself; science, however, represents relatively virgin territory. For me, the triumph of Djerassi's play is to present the language of the laboratory in an involving, accessible way. In our production, complex explanations of infertility were rattled off like the quickfire comic banter in a classic 1940s romantic film comedy and thus, I believe, had more appeal to a broad-based audience not deeply versed in science.

In 1998 when an *An Immaculate Misconception* was first performed, the questions it posed were central to the play's purpose, and they re-surfaced in 2006 in Djerassi's most recently premiered play, *Taboos*.

Taboos. New End Theatre and King's Head Theatre, London, 2006.

If only having a child was as easy as opening a box...

Taboos is a comedy about keeping it in the family; in short, about contemporary sexual behaviour and mores. The play saw Djerassi return to the subject of his first play, namely the very topical matter of assisted reproduction, to which he gave a funny and unexpected twist. (It is perhaps useful, here, to remember that Djerassi was the scientist who 'invented' the Contraceptive Pill, an achievement that puts him in a unique position to reflect on the whole issue of assisted reproduction.)

What do a pioneering West Coast urologist, a celebrated TV presenter, a hip, single San Francisco lawyer and a God-fearing, childless couple from the Southern bible belt have in common? They all want to be parents. Sally and Harriet are a lesbian couple who want to have a child. They believe they can simply make a family by mixing the necessary biological ingredients. All they need is some sperm, and Harriet's brother Max is happy to oblige. Meanwhile, Sally's brother Cameron and his wife Priscilla are desperate to have a baby, having put their faith in God. They just need an egg donor and Harriet is happy to volunteer. But when Harriet decides *she also* wants a baby, and uses sperm from Cameron, a uniquely tangled (and often hilarious) emotional mess results – becoming a biological mother proves to be a rather more emotionally involving process than either couple expects.

What makes one a parent? Love? Genetics? Giving birth? This lies at the centre of Djerassi's play, *Taboos*. Returning to his scientific roots after three non-science plays, he explored the other side of Planned Parenthood and the creation of The Perfect Family. The play explores the unexpected and often messy result that arises when emotions and science collide. Inspired by actual events, the play is about the complex and potentially irreconcilable human and emotional consequences of becoming a biological and/or legal parent, in an era in which social and legal structures have not caught up with scientific advancement – namely the possibilities offered by sperm and embryo donation. The play also examines the conflict between traditional Christianity on the one hand – with its stress on the nuclear family and heterosexual marriage – and on the life cycle and choices of a lesbian couple in San Francisco. Overall the play works to raise questions that are becoming increasingly relevant and controversial in contemporary society.

Taboos was intended as a contribution to the wider debate on human reproduction. The story of an extended family spiralling outwards by unorthodox ways and means has many parallels in the real world. Every month there are new cases that force churches, courts and society at large to come to terms with increasingly wide definitions of parenthood ranging from relatively straightforward in-vitro fertility treatments through heterosexual surrogacies to demands from the gay community to be allowed to be parents. As Djerassi said in the performance programme to the 2006 London production:

Terms such as 'marriage,' 'family,' and 'parent' used to have firm denotations. They were the rock on which our cultural values rested. Terms such as 'embryo,' 'baby,' or 'twin' were also considered unambiguous. Assumptions that marriage must be heterosexual and that a child cannot have two parents of the same sex were never even considered assumptions, because they were beyond questioning. Recently, all these terms have become destabilized, their meanings blurred, their range extended. Some would blame in-vitro fertilization technology during the past three decades for these developments, but in actual fact major social and cultural changes – primarily in the USA and Europe – were even more responsible for the monumental shift that has caused so much fear and antagonism, especially among the ever increasingly strident fundamentalists in the USA. So why not write a play about a situation where 'family' and 'parent' have assumed disturbingly fuzzy meanings?

At the play's close, both 'sides' implicitly acknowledge that their original approaches have not actually served their emotional needs, have not enabled them to understand the needs of others, and may not serve the best interests of the children. Although the ending of the play, with its positing of a new social microcosm, could be seen as a compromise (everyone seems 'happy now'), there is also scope to view it as a radical critique of society itself, and of prevailing ideologies. As the play implicitly suggests, the society in which these protagonists live has either ignored the consequences of science or has actively prohibited any real debate. And we – the people who make up society – are the ones who will eventually suffer.

So, is it science that is pushing these issues forward, or is it society insisting scientists come up with solutions to questions that traditional social institutions can no longer answer? Whatever the response to the question, it is certain that Djerassi's plays deal with issues that are of great significance to us all, and another delightful example of his doing this takes place in the second of his science-in-theatre plays, *Oxygen*.

Oxygen. Riverside Studios, London, 2001; BBC World Service radio drama production, 2001.

Three men discovered oxygen. But only one got there first . . .

Oxygen was co-written with Roald Hoffmann (Frank H. T. Rhodes Professor of Humane Letters, Cornell University). Hoffmann is a Nobel Prize Laureate in Chemistry. The play charts the real-life deep rivalry and controversy in 1777 between the Englishman Joseph Priestley, the Swede Carl Wilhelm Scheele and the Frenchman Antoine Laurent Lavoisier over who would be recognised as having been the first to discover oxygen. Inspired by actual events (and by particular research undertaken by Hoffmann in Cornell University's Lavoisier Collection where he found M. Lavoisier's antique travel chest), the play is about priority, competition, fame and recognition in science, and the moral consequences of these; it is also about the discovery of oxygen and revolutions, chemical and political.

Oxygen alternates between 1777 and 2001, the centenary of the Nobel Prize, when a fictional Nobel Chemistry Committee of the Royal Swedish Academy of Sciences decides to inaugurate a 'RetroNobel' Award for those great discoveries that preceded the establishment of the Nobel Prizes one hundred years earlier. The play poses the question: what if Nobel prizes could be awarded to dead people, to those great experimenters and thinkers of the past who altered the course of science? Such a prize might, at least, be free from the lobbying and feuding that goes on behind the scenes of the annual Nobel Prize-giving ritual, whose prizes are supposed to honor men and women who have made positive contributions to our understanding of the world around us. But is the choice as objective as the science it honors? Written by a pair of multi-award-winning scientists, including a Nobel Prize winner, *Oxygen* also takes a sideways look at the Nobel selection process.

The Committee decides to focus on the discovery of oxygen in the eighteenth century since that event launched the modern chemical revolution. But who should be so honoured? Joseph Priestley, Antoine Lavoisier, and Carl Wilhelm Scheele all independently discovered oxygen, and as a result helped turn 'natural philosophy' into modern science. Scheele probably got there first but he did not tell anyone. Does it matter if you do not tell the world? Priestley also discovered oxygen (then called phlogiston) but clearly did not understand what he had found. Does this matter? So should we acknowledge Lavoisier – who coined the word 'oxygen,' and understood it first – even though he tried to ignore and deceive the others? How will the Committee go about choosing the prizewinner, and why? Then, as now, debates over who discovered what first press every scientist's emotional buttons, revealing them as all too human. This is the fascinating premise behind *Oxygen*. In a fictional encounter, the play brings the three protagonists and their wives to 1777 Stockholm at the invitation of King Gustav III. The question to be resolved: who discovered oxygen?

But what is discovery? And why is it so important to be first? In science, sometimes getting there first seems more important than breathing. The recent furore over who was first to publish the entire human genome has shown that the issues surrounding the all-consuming desire in science to be first are still as controversial as ever. But is it just about patents and money or is there something more fundamentally human at work? Roald Hoffmann said: "People want to understand what has driven scientists, how that precious knowledge has been gained, and whether one has had to sacrifice some aspect of one's humanity to gain it. That's what we address in our play." (Kauffman 40) Importantly, he also said: "One of the things we really need to do in the twenty-first century is humanize science and make it part of world culture. We need to talk to people about science so they can make intelligent, democratic decisions about it." (Rayl 20) *Oxygen* explored uncomfortable truths about scientific discovery, the sometimes unscrupulous competition between scientists, and the dubious business of awarding science prizes. And as with all his science plays, Djerassi wanted to provoke discussion about the ethics of individual scientists. He said: "Our play should not be viewed as a play about history, but about the character and the culture of science and scientists." (Rayl 20)

As a sub-theme (but no less important one) this play also examines the struggle of women trying to break through the glass ceiling of the male-dominated world of science: the man who arguably made the single greatest contribution to women's liberation through the invention of the Pill has ensured that *Oxygen* emphasises the contributions made by women in science.

As with all Djerassi's plays, at the end it did not provide answers; as A.J.S. Rayl said about the play in *The Scientist*: "science has changed the world during the past 200 years, but the scientists, the human beings behind the discoveries, have not." (Rayl 20) Djerassi feels there are no black and white answers, only questions, and that each individual has to engage in the process of answering them. In the case of *Oxygen*, one of the final insights the authors wanted the audience to understand is that the individual scientists in the play, both historical and contemporary, are driven by the same human desires, be they egotistical, selfish or professional, and that in these regards over the years nothing much has changed. Djerassi says: "I like ambiguous endings. We don't want to spoon-feed the audience, and besides, all interesting scientific and social problems are gray. If we can amuse and inspire curiosity that would be wonderful. Sometimes there are no real answers." (Bennett 19)

When working on all the plays, part of the (consistently enjoyable) challenge for me was to find ways to make the science work on stage, both as part of the narrative (story) and theme, and as physical, dramatic events and sequences of stage action. In the case of *Oxygen*, we were able to re-create and reconstruct, live, some of the actual scientific experiments carried out by the seventeenth century scientists featured in the play, even using the type of chemical apparatus the men might have used, as was noted in a press review of the production by Rachel Halliburton in the *Evening Standard* newspaper ("the staged scientific experiments are ravishing"). Whilst there was a suspenseful and fictional plot and sub-plot (much of which was *based* on fact), *all* the science in the play was historically accurate, which made for a highly entertaining and memorable element in the performance. The Royal Institution in London supplied the genuine scientific equipment for the production. It was authentic and historical, beautiful to look at, and added wonderful period detail and atmosphere. Within this universe – a mixture of the 'real' and the 'stylised' – the seventeenth-century scientists and their wives could wander alongside the twenty-first-century scientists (male and female), also allowing the ghosts of the past to observe, comment and intervene.

The mix of historical periods allowed us to have great fun in the designing of the set for the production – the celebrated opera and theatre designer Russell Craig was responsible for a truly beautiful setting which was modelled on Restoration theatre, as noted in the *Evening Standard* newspaper ("a treasure trove of test tubes, flasks, wig stands and a particularly enticing phrenology bust"). He also designed the gorgeous seventeenth-century costumes, and the highly stylish (almost futurist) contemporary costumes. We located the play in a world where a seventeenth-century experimental science laboratory and a twentieth-century Nobel Prize meeting room co-existed, with no sense of where one started and the other ended. The opportunity to explore events which had actually happened in the past allowed us to (re)create colourful stage action, an example of which was the court masque that closed Act 1, which dramatised the battle between phlogiston and oxygen, performed by M. and Madame Lavoisier. The Lavoisiers did actually write and perform such a masque (the music and lyrics for which no longer exist), so we imagined what it might have looked and sounded like. Our composer, the Oscar-nominated Andy Price, researched the music for contemporary court masques and wrote a score which referenced the appropriate historical style. Designer Russell Craig designed sumptuously lovely masks. Djerassi and Hoffmann, in collaboration with poet Lavinia Greenlaw, wrote the amusing lyrics (in verse). Chris Corner designed beautifully atmospheric lighting and Jack Murphy, then a choreographer at the Royal National Theatre, choreographed the movement and dance.

The play, *Oxygen* – and the others which Djerassi and I have worked on together –, ought to bring the language and customs of the scientific world into accessible and exciting dramatic form; in them, we sought to experiment with and find new ways to dramatise science and scientific ideas on stage (at least, to *us*). This was the first theatre play Carl Djerassi and I collaborated on. Memorably (for me), whilst we were planning and presenting the British premiere production (which opened in October, 2001) there were a number of other productions of the play being planned and performed elsewhere in the world: in San Francisco (San Diego Repertory, the world premiere production, directed by Brian Bevell), a German-language production (at the Stadttheater Würzburg, directed by Isabella Gregor), as well as a variety of rehearsed readings, including at Cornell University, in Manhattan, Vienna, Philadelphia, and at the Tricycle Theatre (directed by Erica Whyman) and the Royal Institution in London (the latter reading directed by myself).

At the same time we were also working on a radio drama version of the play, which involved producing a much shorter version of the UK play: a number of the structural changes we made for the radio version subsequently found their way into the theatre version. Finally, the American publishers of the play were rushing a version of the text into print (which meant that many scriptural features of the UK production were not published in the book version as it was printed, unusually, before the play's world premiere). So many different companies simultaneously presenting readings and productions of one of Djerassi's plays, in various locations around the world, never occurred again in quite the same way, but it did make for a very energised and exciting time – I well remember the authors sending the same emails to *all* the four *Oxygen* directors working on the play! The readings, radio version and overseas productions – and the version of the play that ended up being successfully published – usefully fed into the dramaturgical work we did on the play in readiness for our UK theatre production. In fact, the wonderfully talented Austrian director Isabella Gregor and I regularly discussed the work we were doing with Djerassi and Roald Hoffmann, often sharing insights into how the play might be developed and sharing new versions of scenes. I am delighted to say the relationship Isabella and I began with this play has continued over the years, and we have frequently discussed work we have been doing with Djerassi on his later plays.

Calculus. New End Theatre, London, 2004.

Sir Isaac Newton. He saw the light. But lived in darkness ...

Calculus, a play which explored the behaviour of one of the greatest male scientists of all time, Isaac Newton, is a favourite of my own amongst the productions of Djerassi's plays I have directed. It underwent the most dramatic of transformations from first draft (in 2001/2) to the one we eventually performed in London (in 2004). As was the case with all of Djerassi's plays I worked on, I would go to him with director's suggestions about how I might 'bring off' a particular theatrical, dramatic effect, or how I was visualising a particular sequence or wished to physicalise or stage a particular idea, and this sometimes resulted in Djerassi re-writing the script to accommodate those directorial ideas. For example, a year or two later, after we had produced a staged rehearsed reading of the play before an audience, I had an idea of how I wanted to stage the play, and made a proposal for a radical re-locating of the action, and for new scenes, which Djerassi was happy to take on board. In due course, characters were cut from the script and the setting for the play moved from anonymous club-rooms in seventeenth century London to a rehearsal room-cum-office in the attic spaces of the Drury Lane Theatre. As two of the leading characters in the play were real-life men of the Restoration Theatre – playwright-architect Sir John Vanbrugh (who designed and oversaw the building of London's Haymarket Theatre and Blenheim Palace) and the then actor-manager-playwright of Drury Lane Theatre, Colley Cibber – the suggested re-location seemed to make sense and hopefully gave the play/production more resonance. The idea to re-locate the play was actually inspired by Djerassi's use of the classic device of a play-within-a-play, through which *Calculus* explores Newton's abuse of his position by his ruthless manipulation of the people around him. The play-within-a-play was used "to denounce the Royal Society's deliberations as no more than a drama written by Newton himself." (Campos 12) Later changes to the script included adding another twelve characters in a new scene which featured all the members of the committee, a feature of the production that one press reviewer, John Thaxter in *The Stage*, commented on saying that no other production in that particular theatre had ever had so many actors in it ("a Restoration-style comedy that brings on the stage the largest cast ever assembled at the New End Theatre").

Sir Isaac Newton, England's foremost scientist, has accused the German polymath, Gottfried Wilhelm Leibniz, of scientific plagiarism. Scandal ensues. In 1712 the Royal Society establishes an anonymous commission of eleven good men to adjudicate. But has their decision already been taken for them elsewhere? The play draws us into the dilemma: prostitute their reputations in order to please the all-powerful Newton or be true to their conscience and reject his manipulations? This vividly theatrical play offers an insight into the all-too-human side of the man recently voted Man of the Millennium by *The Sunday Times* newspaper.

Based on actual events, and written in a style reminiscent of the theatre of the time, *Calculus* has intrigue, wit, colourful characters, and a delightfully surprising twist in its tail. The play is about Sir Isaac Newton and his accusation of scientific plagiarism levelled at German polymath Gottfried Wilhelm Leibniz. Acknowledged the world over as perhaps the UK's greatest scientist whose work on gravity and optics ushered in the Age of Scientific Reason, Newton was also remote, secretive, puritanical and vindictive, with questionable ethics, who abused his position of power and trust for his own self aggrandisement. *Calculus* offers a unique opportunity to experience a powerful view of this deeply flawed genius, providing a compelling and revealing picture of this towering figure of history while exploring differing aspects of his obsessively competitive nature. In parallel, *Calculus* explores the abuse of power in high places and the establishment of a supposedly 'independent' investigation to clear the accused. If this seems familiar, then that is no coincidence. The scandal that took place 300 years ago may be obscure now but the play's themes have an alarmingly contemporary resonance about them. The play also concerns itself with the thorny issue of precedence in science: the importance of being accepted as the first person to discover or invent something (a theme shared with *Oxygen*). As the work of scientists plays an increasingly intrusive role in all our lives, and patents on new discoveries become ever more valuable, this issue is set to become a major concern for society to address.

Calculus is a mathematical method of defining the mechanics of moving objects, crucial to the development of much of modern technology. It was invented independently by Isaac Newton and Gottfried Wilhelm Leibniz, one of Germany's greatest eighteenth century scientists. Rather than dealing with their competing theories, the play addresses the *moral* calculus arising from the decades long struggle between Newton and Leibniz over who invented this vital bit of mathematics first. *Calculus* is concerned with illuminating the personal ethics and behaviour of scientists rather than their science:

Calculus deals more with scientists than science; more with scientific ethics and behaviour than scientific facts....It is through the story of some of Newton's 'little squires' that Calculus tries to examine one of Newton's greatest ethical lapses." (programme for the 2005 San Francisco production)

A character in the play asks: "What purpose is served by showing that England's greatest natural philosopher is flawed like other mortals? We need unsullied heroes!" (Djerassi 2003, 171) But what if the hero *is* sullied? What if his personal ethics are not as pure and objective as his science? How should we view his achievements then? *Calculus* – as did *An Immaculate Misconception* and *Oxygen* – dramatises the ways in which the ethical failures of scientists can interfere with the science they do. The play explores an issue – what does moral integrity have to do with integral calculus? – as relevant now as it was 300 years ago; that the personal ethics of a scientist can never be divorced from the science they promote. In this way, similar to Tom Stoppard's *Arcadia* and Michael Frayn's *Copenhagen*, science is used as a metaphor:

Calculus refers not only to maths, but to the character's moral and political calculations. In fact, the play contains very little maths, with the exception of a comical scene in which de Moivre illustrates calculus by eating an apple (of course) at varying speeds and then calculates his velocity at a given point in time. (Campos 12)

The question of metaphor in relation to Djerassi's play is an intriguing one. Indeed there is some debate amongst critics about the use of metaphor in science plays in general (there are too many or there are not enough!). Djerassi once said when speaking about his play *Oxygen*, it should be viewed on its own terms, "not as a metaphorical play but a factual one." (Lynch 22) For Djerassi, it is important that science in his plays is "not merely a metaphor, but the narrative heart and soul." (Zehelein 105) For me, however, drama gains deep resonance through the use of metaphor, whether this is conjured through the writing or through the way the play is directed, designed and staged (the *mis-en-scène*). This question therefore became another balancing act in the developing of each play. In the case of *Calculus*, the brilliantly designed and realised set – the work of the gifted Michael Taylor, who also designed the clever and gorgeous sets for *Phallacy* and *Taboos* – was itself a visual metaphor for the play's themes and its 'world.'

As we know, Djerassi is honest about the pedagogical function of his plays. He has also, in his many fascinating writings about what he describes as *science-in-theatre,* stated that what characterises plays that simply *refer* to science or a scientist in their plays (as opposed to making science or scientists absolutely *central and fundamental to them*) is the use of metaphor, where the play(s) use scientific ideas or concepts and principles to make a metaphorical point. He is right. But from my perspective as a theatre practitioner – and a thoroughly pragmatic one – I am perfectly delighted when writers are able to create metaphors of any type because they have the capacity to elevate the play and its themes in such a way that it becomes more poetic, powerful and memorable. The reality is – whatever Djerassi might have said in the past – that his plays also (and pleasingly) contain metaphor in these ways. In this context, metaphors can be used to make ideas and concepts more accessible to the audience, something I was always conscious of seeking to do when directing Djerassi's plays.

Calculus completed the trilogy of Djerassi's science-in-theatre plays. As with *Oxygen* before it, part of the dramaturgical development of the play involved presenting staged rehearsed readings of the play before an audience, followed by a public discussion with the writer, director and actors. This process of testing the script in front of an audience, and collecting feedback, can be immensely helpful to playwrights as they continue to modify the text and visualise how it might work in performance. This approach was also crucial to the way the script of *Phallacy* evolved over an eighteen-month period.

Phallacy. New End Theatre and King's Head Theatre, London, 2005.

How far will each go to prove themselves right and the other wrong?

A revered classical statue, for centuries thought to be a Roman original, is suddenly proved to be merely a sixteenth-century cast. But is it worth less? And has it become less beautiful? Once again inspired by actual events (this time in the art world rather than from science), Djerassi created a story of suspense and intrigue in this highly sophisticated and entertaining play in which personal rivalries and professional reputations clash as Art and Science join battle in a sharply topical play, a comedy about academic manners and academic in-fighting. One of the central issues in the play was summed up by a critic in this way: "the fascinating main issue is whether a work of art that is aesthetically pleasing becomes more so the more ancient that we believe it is." (Hickman 2005) Intriguingly, the play also examines "the relationship between financial worth and beauty" (Neill) and raises "questions about how we define and value art" (Berkowitz).

She is a top art historian in a world-class museum, he is a professor of chemistry brought in to date a priceless artwork. She is in pursuit of artistic truth, he scientific truth. But can these two cultures ever be compatible? And how far will each go to prove themselves right and the other wrong?

Phallacy references the "The Two Cultures" debate, Art and Science, kick-started by C. P. Snow's Cambridge lecture paper, partly because Carl Djerassi himself traverses both disciplines thus making rivalry between art and science a very suitable subject for him. Furthermore, Djerassi knows a bit about art, too: "... as a serious art collector himself, he is also well situated to examine the aesthetic as well as scientific side." (Rohn) It has even been suggested, perhaps correctly, that the two main characters in the play, the art historian ("the artist") and the scientist, represent parts of the author himself. The art historian believes in beauty before truth, the scientist believes in truth before beauty and is contemptuous of a profession that does not give primacy to facts. "Djerassi's point is clear: you don't always need high-tech science to demonstrate the phallacies of others' arguments." (Milgrom 10) At the heart of this play there was a plea that the two tribes make some accommodation, one with the other.

For myself, I have observed that scientists use as much 'creativity' in the pursuit of their aims as the artists do in theirs, so I suspect there is less of a chasm between the two cultures than there perhaps was. And even though – to paraphrase Djerassi – on the face of it, science is about concrete experiment and the arts are about applying the imagination, I suspect that what motivates people to do any creative work is more alike than we care to own up to. Roald Hoffmann, co-writer of *Oxygen* and author of several non-fiction books dealing with science, literature and the arts said, very pertinently: "Science can provide creative and spiritual satisfaction, much like art, music and literature, and people are interested when the emotions aroused by these disparate elements are combined." (Bennett 19) I have personally been delighted to see an ever-increasing number of interdisciplinary collaborations between scientists and artists, where practitioners of both disciplines are seeking out areas of common interest, enquiring into ways in which the human spirit in all its majesty can be better and more deeply explored through scientific and artistic contemplation, discovery and experimentation. In discussing *Oxygen*, Suzanne Lynch in *The Irish Times* (2001) said:

The play also suggests ways in which the domains of humanities and science can overlap
.... It marks a bold attempt on the part of science to take its place in the world of the humanities – and to exploit the potential of theatre to bring scientific knowledge to a wider audience. In this sense, *Oxygen* is an important stage in the move towards a more inclusive form of education and pays testimony to the power of theatre to open up the possibility of an interdisciplinary way of viewing the world.

This remark is as true of *Phallacy* as it is of the rest of Djerassi's plays. It is pleasing to me to know that Djerassi and I, in regularly working together on his plays, are conducting an on-going interdisciplinary investigation, where a scientist collaborates with a theatre director.

With some of the plays – *Oxygen, Calculus, Ego, Three on a Couch* – in order to get the very best, in dramatic and theatrical terms, out of Carl Djerassi's strikingly original ideas, to identify striking and imaginative ways of staging and bringing the plays to theatrical life, the scripts sometimes went through a number of drafts. *Phallacy* underwent extensive change from first draft stage to when it played before an audience. For a start, there were three different titles, beginning with *Sham*, then *Prick* (as in 'penis,' a fixture highly relevant to the play's plot), and finally *Phallacy* (which is a conflation of the words phallic and fallacy, both very significant components in the play). The play also had a wholly different opening from the one envisaged in the first draft. In the production that played in London, I created a form of prologue, but one without words, one that was highly visual, choreographed, and with a musical underscore. This stylised opening sequence to the show blended the different historical periods in which the play was set and acted as a form of metaphorical prologue. This sequence then dissolved into an animated modern-day audio-visual lecture given by the leading female character, the art historian, Regina, a scene that made extensive use of pre-recorded video footage, a multimedia device which most of my productions of Djerassi's plays have featured as a kind of stylistic signature.

In respect to my work as a director of the plays, I had an opportunity to create productions which synthesised my two personal theatrical pre-occupations, namely a keen interest in text and how to elucidate its meanings, and theatre that is less concerned with text and words, a type of theatre (also known as *performance*) which mobilises and seeks to integrate inter- and multi-disciplinary approaches. Both types of theatre are concerned with ideas, but the first is more pre-occupied with psychological realism, seeking to make a profound connection on an intellectual and emotional level with its audience. The second is the theatre of images, of visual and physical theatre, theatre that mixes media, which creates a new hybrid of visual imagery, movement, sound, language, scenography, a theatre which transmits it meanings as much through the senses and through its own artifice as it does through the cerebral context.

I am conscious that there may be something rather 'British' about this seeming desire for balance between textual and visual approaches. The fact is that I would love to have the freedom to use more 'European' approaches to my directing work, where the British tradition of textual veracity and dependence on words and language can sometimes feel like a cultural and creative straightjacket. That said, there are a number of historical, political and cultural explanations as to why different traditions have evolved between Britain and the European mainland and, increasingly, there is more and more cultural and creative convergence with approaches to theatre throughout Europe. In Britain, even mainstream theatre is being profoundly influenced by the more iconoclastic and visual styles of European theatrical practice and, in terms of playwriting, British dramatists continue to have great success throughout the rest of Europe.

Here I think it is important to note that Djerassi's plays were not wholly or straightforwardly naturalistic or realistic (even though they were written in the 'classic mode' of 'the well-made play'). This factor was of interest to me on a number of fronts. Firstly, in line with a number of other science plays, Djerassi's plays were avowedly text-driven, where ideas, themes, words and language were majorly important, a fact I had always to be conscious of as the director. But because of my interest in the more performative aspects of the process, there was, for me, bound to be an important relationship between text and performance or, if you will, between the play as text and the play as a performance. I do not wish here to be drawn into questions about dramatic performance in relation to meaning or understanding of a text or about the primacy of performance over text (though very interesting), but I should state for the record that the dramatic text, in the case of Djerassi's scripts, remained constant and (to use Eva-Sabine Zehelein's phrase) "the only durable element in every performance." (Zehelein, 79) Zehelein correctly goes on to say that

(in many) science plays ... it is the text itself, and not the performance, which reveals the differences between science and/or scientists and their respective (metaphorical) functions ... plays which take up socio-culturally or socio-politically relevant topics and wish to foster public discourse, revert to classic formats and put the text at center, since these formats are much more accessible for a theater audience .... Science plays are, indeed, in many cases a means of communication more than an artistic medium for its own sake .... (Zehelein 80)

This is very true of Djerassi's plays, at least from the perspective of the writer, and is probably part of their appeal to audiences. However, from my perspective as director-dramaturg, in production/performance the play ought to take on another life; as Zehelein says, "(the) dramatic text, of course, develops its own dynamics once it is turned into a theatrical production, and both, text and production, are cultural artifacts in their own right." (Zehelein 81)

As a director, I want the writers I work with to know that I believe in 'writer's theatre' (which I do); in return it is my hope the writer will believe in 'director's theatre' (which I hope they do). Even though it appears most science plays are heavily dialogue-dependent – which includes Djerassi's – and need text and the spoken word to convey their meanings, and that this is the main focus, I nonetheless worked hard as the director to 'add performative value' to the play in the way in which I staged it. I am reminded here of what the great director Michael Blakemore said about his wonderful production of Michael Frayn's play *Copenhagen* (Blakemore has directed most of Frayn's plays): "Putting on a play itself is a sort of scientific experiment .... This is not a naturalistic play. We're not trying to pretend that what we're seeing is *real*. The audience must listen to the arguments, empathize with the character's emotions, and create the reality for themselves." (quoted from Shepherd-Barr 104) This is what I aspired to in the directing/producing of Djerassi's plays, but I may not be the appropriate person to say whether I ever achieved it.

My productions sought to be contemporary in look and feel, even though most of the plays had historical components to them. It was important for me to create a world in which the historical past co-existed with the modern, twenty-first-century world I wanted the audience to see and experience. To this end, I developed a visual and aesthetic approach which overlapped both worlds, or many worlds or modes of existence: the corporeal past with the 'real-life' present, alongside the ghosts of the past, to create a stylised, highly theatrical and poetic stage-world where historical characters could co-exist with present-day characters, where the ghosts of the past could visit, observe and interact with the present-day characters. All the plays allowed me to explore these visual and physical ideas in the way I staged them and, obligingly, Carl Djerassi worked with me so that the productions could incorporate these sorts of stylistic approaches.

In practical terms, it was important to find interesting and innovative ways of illuminating the relationship between the past and the present, which went beyond a (too) literal staging, for example, physically separating the historical characters from the modern day characters on either side of the stage, or by simply having higher and lower stages; I much prefer to find ways of having the past merging with the present, co-existing in the same space, as was noted by Heather Neill of *WhatsOnStage.com* about my production of *Phallacy*: "Andy Jordan's fluid production in which characters move around each other, perhaps centuries apart, without causing confusion." For me this feels and looks like a filmic style, where physical action as it relates to the various characters on stage can dissolve from one plane to another, one dimension to another, where we are not always aware of the joins or, alternatively, we can instantly cut to another sequence, where we *are* profoundly and joltingly aware of the joins. My work in theatre (and in radio drama) is strongly influenced by film and its vocabulary, and by the extraordinarily intense emotional power of music.

My work as a director – in whichever medium I am working – invariably starts with my need to get a sense of the 'music' of the play, conjured up by the rhythms and tempi in the writing, and from the world of the play. When starting my work it is of equal importance to imagine the look of the play, and I most often begin this process by identifying a painter or paintings that seem to capture or represent the visual style and feel of the play and/or production I am envisaging. Getting a clear sense of these two components – the visual style of the production, and the sound of it – is a vital part of my creative process as a director. In the case of theatre, I am fascinated by the power of the image, of stage pictures, by the shapes made in space and time, and by using the visuals as a structural and metaphorical storytelling device. I also delight in the way that movement on stage – be it that of the actors and their physicality, the dramatic and physical action, or the structure and flow of the play or performance – can become of itself visually transfixing, haunting and potent, like a ritualistic or sensual dance, so that it is not always linear and literal but plays around with time and space, order and consequence, and thus becomes more powerful and striking by being poetic and allusive.

In summarizing my contribution to Carl Djerassi's plays, as director-dramaturg, I would say it is that I have attempted to blend contemporary performance trends with his dialogue-based approach. Being granted such a regular chance to explore these directorial approaches is a great privilege, as I am deeply fascinated by both text-based theatre and by non-text driven styles. As I write this I am looking forward to directing and producing the UK productions of Carl Djerassi's next two theatre plays, *Foreplay* and *Insufficiency*, which will be Djerassi's eighth and ninth plays. Djerassi's enthusiasm, energy and ambition is remarkable and inspirational. How many other playwrights in their eighties are so ambitious and prolific? I am enormously grateful to Carl Djerassi for the opportunity he has given me to work on a still growing body of work, and to have done so with one of the truly most extraordinary men I have ever had the great pleasure and good fortune of knowing.

Works Cited

Bennett, Dan. 2001. "Experiments in Human Nature." *Los Angeles Times* 25 March: 19.
Berkowitz, Gerald. 2005. "Play review; Phallacy." *Theatreguide London* (internet magazine) 18 April. Quoted from www.djerassi.com/phallacy/reviewquotes.html
Campos, Liliane. 2004. "Examining Newton's darker side." *PhysicsWeb* 9 August: 12.
Devins, Dorian. 2001. "Discovering Oxygen." *The Dramatist* (New York) 4 March: 28.
DJANUS. 2005. *Calculus*: programme. San Francisco: San Francisco Performing Arts Library & Museum.
Djerassi, Carl. 2001. *This Man's Pill: Reflections on the 50th Birthday of the Pill*. Oxford: Oxford UP.
Djerassi, Carl. 2003. *Newton's Darkness: Two Dramatic Views*. London: Imperial College Press.
Djerassi, Carl. 2007. "When is 'Science on Stage' really Science?" *American Theatre* 24, 28 January: 96-103.
Halliburton, Rachel. 2001. "Flirting gets Scientific." *Evening Standard* 20 November: 44.
Hickman, Julia. 2005. "Play review; Phallacy." *Theatreworld Internet Magazine* 16 April.
Hughes, Kevin. 1981. "Play review; Theory for the Attention of Mr Einstein." *Event* (London) 21 August: 37.

Kauffman, George B. & Laurie N. Kauffman. 2003. "Oxygen: A Play in Two Acts." *The Chemical Educator* 8/2, 30 June: 40.

Lynch, Suzanne. 2001. "A winning formula for the stage?" *The Irish Times* 23 November: 22.

Milgrom, Lionel. 2005. "Play review; Phallacy." *Chemistry World* 17 April: 10.

Neill, Heather. 2005. "Play review; Phallacy." *What's on Stage Internet Magazine* 18 April. www.whatsonstage.com/reviews/theatre/london/E8821113834766/Phallacy.html

Rayl, A.J.S. 2001. "Putting a Human face on Science – Renowned chemists advance science through the arts." *The Scientist* 15, 16 October: 20.

Rohn, Jennifer. 2005. "Science-in-theatre: a chemist attempts to translate." *LabLit* (internet magazine) 30 March. www.lablit.com/article/21

Shepherd-Barr, Kristin. 2006. *Science on Stage: From Doctor Faustus to Copenhagen*, Princeton: Princeton UP.

Thaxter, John. 2004. "Play review: Calculus." *The Stage.* 5 August: 9.

Tyler, Christian. 1999. "Qualified Opinions on a Matter of Life and Death." *Financial Times* 15 August. 14.

Zehelein, Eva-Sabine. 2009. *Science: Dramatic: Science Plays in America and Great Britain, 1990-2007*. Heidelberg: Winter.

Reviews

An Immaculate Misconception, Bridewell Theatre, London, 2002

UK Production History: The play was premiered at the 1998 Edinburgh Festival and performed in London in 1999 at the New End Theatre, both to considerable acclaim. Andy Jordan acted as Consultant Producer to the original 1998 Edinburgh Festival production. He then directed a BBC World Service radio production in 2000 starring the leading British theatre, television, film and radio actors Henry Goodman and Penny Downie. The play has also been broadcast by German, Swedish, US and Czech radio. Finally, Andy Jordan produced and directed a revised version of the play in London in 2002 at the Bridewell Theatre. Subsequent productions took place in quick succession in cities including San Francisco, New York, Vienna, Cologne, Munich, Berlin, Sundsvall, Stockholm, Sofia, Geneva, Tokyo, Seoul, Los Angeles, Lisbon, Singapore, Detroit, and Zurich.

Cast: Paul Moriarty, Debra Stephenson and Terence Wilton. Directed by Andy Jordan, Designed by Joanna Parker, Video Design by Nic Sandiland, Music by Iain Dunnett.

"Few people will ever have thought that a sperm fertilising an egg could make a great spectator sport. But in Carl Djerassi's up-to-the-nanosecond comedy, it becomes edge of the seat, whoop-worthy viewing .... In a week when Germaine Greer has asked, 'Do we really need men?,' on the grounds that the billions of sperm they produce mean that women could theoretically conceive if '99.9 percent of human males were wiped out,' Djerassi's drama flings up a thousand pertinent questions. As the chemist who synthesised the Pill, his exploration of a world where sex and reproduction could become separate is satisfyingly informative, as well as containing some distinctly Billy Wilder moments ....

But the entertaining presenting of science on Joanna Parker's set – the epitome of laboratory chic – makes this well worth viewing ...." (Rachel Halliburton, *London Evening Standard*)

"Science is the new art form, states a character in Carl Djerassi's play ... but can it make for good theatre? Happily, the answer is yes, thanks to a cunning combination of technology with a tender tale of love and rivalry between scientists at the edge of discovery .... Not least the pleasures of Andy Jordan's production is the combination of on-stage action with microscopic images of a real-life ICSI .... I won't spoil your pleasure by revealing the final outcome, which keeps you guessing to the last moment." (John Thaxter, *The Stage & What's On in London*)

"A couple of hundred years ago, science used to be an art form, and also an entertainment. A leading scientist would perform an experiment in the presence of a breathless audience of adults and children – some of these events have been recorded by artists who have clearly captured the sense of occasion and wonder. With the increasing specialization of science and scientists we gradually lost the drama and creativity of it all .... So I was glad to see this fascinating play, which is about the relentless march of science and its effects on human reproduction. Oh, and its effect on real human beings too. You and me .... The set, designed by Joanna Parker, made great use of the space, with racks full of brightly coloured potions, and a screen onto which was projected films of actual ICSI fertilizations. Directed and produced by Andy Jordan this is an enjoyable and informative evening which will get you pondering some of the most intractable questions of our time." (Julia Hickman, *Theatreworld Internet Magazine*)

OXYGEN, Riverside Studios, London, 2001

UK Production History: The play was given a staged rehearsed reading in 2000 as the first-ever presentation of a theatre play at the Royal Institution in London, in the historic Michael Faraday Lecture Theatre, hosted by the RI's then director, Dame Susan Greenfield, and directed by Andy Jordan. The play opened at the Riverside Studios in London in 2001, directed and produced by Andy Jordan, with a cast of renowned British actors. A BBC World Service radio drama production, directed by Andy Jordan, aired to 36 million people followed in 2002, featuring equally eminent UK actors such as Miles Anderson and Jan Ravens. Subsequent productions took place in quick succession in cities including Wellington, New Zealand, Korea (Pohang and Seoul), Tokyo, Toronto, Madison, WI, Columbus, OH, Ottawa, Bologna, Sofia, Glasgow, Porto, Medellín, Rio de Janeiro and São Paulo as well as many other German and American venues.

Cast: Jack Klaff, Catherine Cusack, Geraldine Fitzgerald, Paul Goodwin, Lucy Davenport and Robert Demeger. Directed by Andy Jordan, Designed by Russell Craig, Choreography by Jack Murphy, Music by Andy Price.

"A thoroughly engrossing evening, under Andy Jordan's imaginative direction." (*The Stage*)

"Combining sex, gender politics, academic rivalry and the odd clockwork mouse, they ignite enthusiasm in the most unscientific of minds .... A lively, flirtatious presentation of science .... the oxygen masque and the scientific experiments are ravishing." (Rachel Halliburton, *London Evening Standard*)

"... director Andy Jordan neatly highlights the fact that 18th-century boffins were as competitive and kudos-hungry as their modern counterparts ... an enjoyably laudable attempt to demystify science." (Ian Johns, *The Times*)

"In Andy Jordan's astutely directed production .... The play informs and entertains in equal measure ... a provocative and insightful evening ...." (Lucy Popescu, *Theatreworld Internet Magazine*)

CALCULUS, New End Theatre, London, 2004

UK Production History: The play was given staged rehearsed readings in 2001 at the Theatre Royal in Oxford and, in 2002, at the Royal Institution in London, in the historic Michael Faraday Lecture Theatre, both directed by Andy Jordan. The play opened at the New End Theatre in London in 2001, directed and produced by Andy Jordan, with a cast of highly reputable British actors. There have been various other productions of the play around the world, and an original opera (composed by Werner Schulze) based on the play was premiered at the Zurich Opera Studiobühne in May 2005. The play has just opened (and been published) in Coimbra, Portugal (2011).

Cast: Lynette Edwards, Michael Fenner, David Gant, John Kane, Roger May, Susan Sheridan and Nick Wilton. Directed by Andy Jordan, Designed by Michael Taylor, Dramaturgy by Darren Tunstall and Andy Jordan.

"*Calculus* is brilliantly theatrical ... there's more than enough sparkling originality to keep an audience entertained .... this is daring stuff ... a bold and stimulating drama whose style ... is utterly unique." (Helena Thompson, *Ham & High*)

"... a Restoration style comedy that brings on the stage the largest cast ever assembled at the New End Theatre ... sumptously designed by Michael Taylor ... the evening is carried off in triumphant style ... " (John Thaxter, *What's On in London*)

"Andy Jordan's production has much to recommend it, not least the playfully theatrical set design by Michael Taylor. There are imaginative flourishes throughout, in particular when Michael Fenner's Cibber and David Gant's Vanbrugh transform themselves into Leibniz and Newton...." (Robert Shore, *London Time Out*)

"Carl Djerassi completes his Science in Theatre trilogy with this witty unravelling of the motives behind (the question): which came first – differential or integral calculus? ... The setting is actor-manager Colly Cibber's dressing room at Drury Lane." (John Thaxter, *The Stage*)

PHALLACY, New End Theatre, London April 2005, transferring to the Kings Head Theatre, May 2005

UK Production History. The play opened at the New End Theatre in London in 2005, directed and produced by Andy Jordan, with a cast of highly reputable British actors.

There have been various other productions of the play around the world, including a German radio version broadcast in 2006, the New York premiere in 2006 at the Cherry Lane Theatre and the Portugese premiere in Porto in 2011.

Cast: Hamish Clark, Chris Brazier, Karen Archer, Lynette Edwards, Lucy Liemann, Josh Cohen, Jo Bending and Jack Klaff. Directed by Andy Jordan and Warren Hooper, Designed by Michael Taylor, Music and Sound by Jon Nicholls, Dramaturgy by Darren Tunstall and Andy Jordan.

"... an elegantly crafted drama about the conflicts of art and science .... bracingly acted by Karen Archer and Jack Klaff respectively." (Mark Shenton, *Sunday Express*)

"Andy Jordan's lively, slick production brings out the best in the play .... A tautly directed, entertaining production of a thought-provoking play. The fascinating main issue is whether a work of art that is aesthetically pleasing, becomes more so the more ancient that we believe it is .... this is intriguing and thought-provoking stuff, and very well acted." (Julia Hickman, *Theatreworld Internet Magazine*)

"Part detective story, part satire of academic infighting, the play rolls along enjoyably under Andy Jordan's fluid direction, comfortably mixing broad comedy with thought-provoking debate." (Gerald Berkowitz, *The Stage*)

TABOOS, New End Theatre, London, 2006

UK Production History: The UK premiere of Carl Djerassi's Taboos *was presented at the New End Theatre in London in 2006, directed and produced by Andy Jordan. It was subsequently produced at the Theater-im-Bahnhof in Graz, Austria, 2007, and in New York at The Soho Playhouse, 2008. There have been various other productions of the play around the world.*

Cast: Nicola Bryant, Jane Perry, Kathryn Akin, James Albrecht and Jeremy Lindsay Taylor. Directed by Andy Jordan, designed by Michael Taylor, Music and Sound by Jon Nicholls, Dramaturgy by Darren Tunstall and Andy Jordan.

"This is a MUST see play, by far one of the best performances I've seen, and another major triumph for everyone involved .... You'll definitely walk away thinking about where you stand on the issues it raises." (Marcela Olivares, *Indie LONDON*)

"A play that's worth seeing because it explores complex moral and ethical dilemmas – many of which may seem insoluble." (Peter Brown, *London Theatre*)

"Entertaining and thought-provoking play. We may think we are ethics-savvy in this day and age, but the reality may be something else entirely. The bestowal of this awareness, in the end, is one of the best things about *Taboos*." (Jennifer Rohn, *LabLit*)

"Having grappled with such issues as test-tube reproduction and scientific fraud, Djerassi now explores the implications of modern fertility techniques or, as one character describes it, 'a spectacularly complicated reproductive mess among five adults, and not all of them consenting.'" (Ian Johns, *The Times*)

"…a stage full of scientific and ethical viewpoints …. Djerassi manages to make this comedy of genetic muddle gripping …… director Andy Jordan keeps this post-nuclear family buzzing along." (Fiona Mountford, *London Evening Standard*)

"*Taboos* really couldn't be more current …. The changes we are witnessing in Djerassi's 'sex in an age of mechanical reproduction' throws up serious ethical dilemmas, but they also offer exciting opportunities to renegotiate the social and political networks in which we operate." (Helen Birtwistle, *Culture Wars*)

An Immaculate Misconception, 2000, leaflet design
©Debra Hubball

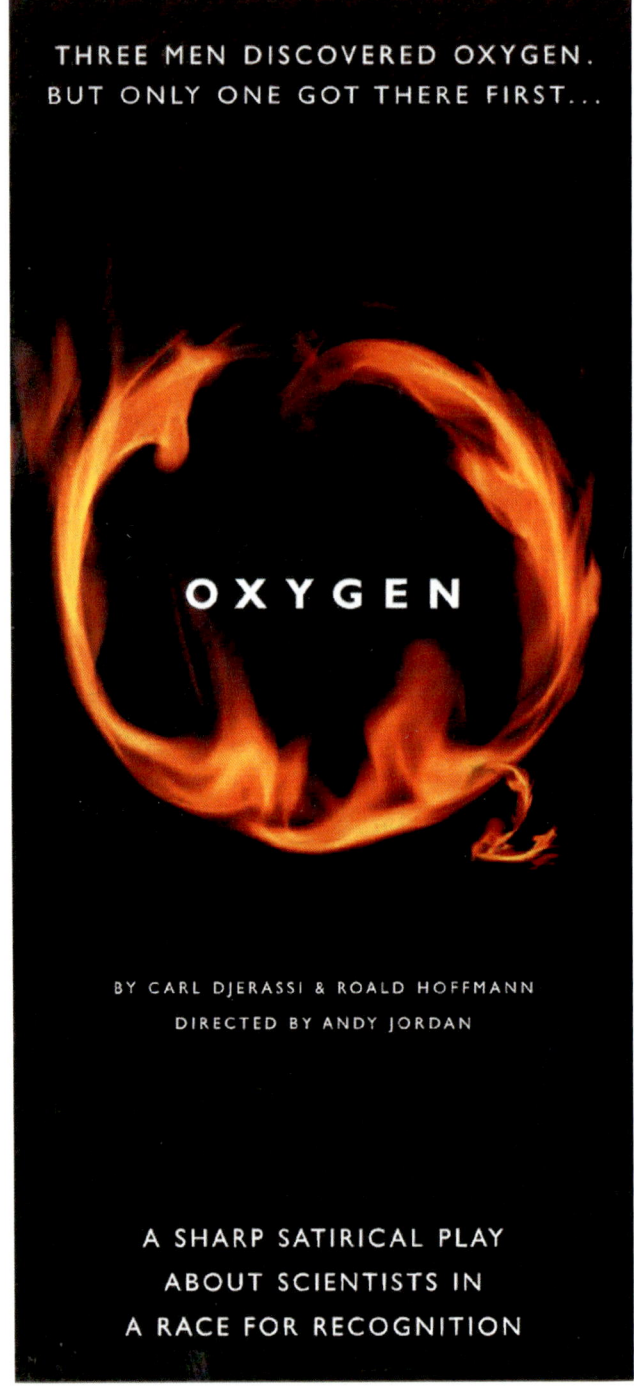

Oxygen, 2001, leaflet design
©Annie Rushton

SCIENCE-IN-THEATRE PLAYS: THE THEATRICAL REALIZATION

ANDY JORDAN PRODUCTIONS

CALCULUS
BY CARL DJERASSI

SIR ISAAC NEWTON.

HE SAW THE LIGHT.

BUT LIVED IN DARKNESS.

Calculus, 2004, leaflet design
©Debra Huball

Phallacy, 2005, leaflet design
©Debra Huball

Taboos, 2006, leaflet design
©Debra Huball

Science-in-Fiction

"If something like a next life exists, yes, I'd rather be a woman": Carl Djerassi's Science-in-Fiction and the Regendering of the Natural Sciences

Ingrid Gehrke

In a 1999 interview I asked Carl Djerassi about his preference for female characters in his novels. He replied that women's lives and careers are simply more exciting than men's. I then asked him whether – in another life – he would prefer being a woman. His answer was:

> If something like a next life exists, yes, I'd rather be a woman, because I am curious and of course there is also a sexual component to this. I also think it should be very interesting ... since many barriers and obstacles which have existed for women so far have been overcome, at least in theory. I'd love to be one of the women who will overcome these barriers first in science or in politics. I imagine this to be very exciting. (Gomboz 9)

In his current life, Carl Djerassi used two opportunities to integrate a female perspective into his work: first as a teacher interested in feminism and then as a writer who likes to create female protagonists. The present contribution focuses on the presentation of male and female scientists in Djerassi's "science-in-fiction" novels, even though his portrait of science as an academic discipline appears to be mainly a story about men. It also tries to answer the question whether the dominance of men and a masculine style in science has historical traits by referring to Londa Schiebinger's research.

Most of Carl Djerassi's literary texts focus on the norms and values of science. In the afterword to his first "science-in-fiction" novel, *Cantor's Dilemma,* he introduces the term "culture" in relation to science and characterizes it in greater detail: "Publications, priorities, the order of the authors, the choice of the journal, the collegiality and the brutal competition [...] are soul and baggage of contemporary science." (Djerassi 1989, 230) In all of his four "science-in-fiction" novels, the leading representatives of this academic culture are men. Whereas in *Menachem's Seed* and *No* ethical issues and the applied field of science in the biotech industry are the main points of interest, *Cantor's Dilemma* and, to a lesser extent, *The Bourbaki Gambit* take a critical look at science in the context of the university.

The protagonist in *Cantor's Dilemma* is Isidore Cantor, nick-named I.C. He is the male representative of the culture described in greatest detail and can therefore be seen as Djerassi's prototype of a scientist. The male protagonists in *The Bourbaki Gambit*, Max, Sepp, and Hiroshi, are either about to retire or have already finished their academic career, whereas Cantor is still a very active and successful member of the academic community. He leads a fairly large research group, he publishes in the best journals, he is recognized and acknowledged by his academic peers, but he is not at an ivy-league university and still lacks the Nobel Prize – the ultimate goal of every scientist – in his award gallery. Cantor is described as a hard working, competitive, ambitious, focused, egocentric, power-conscious and strategically thinking man, who seems to have practically no private life at all. But his favorite student, Jeremiah Stafford, describes him as an exception, because he still spends time in the lab and takes his role as a mentor seriously – at least so Stafford thinks at the beginning of the novel.

All seems to take a very different turn when Cantor has the idea of his life – he scribbles it on the back of a laundry list in the middle of the night so as not to forget it. His theoretical concept needs to be proven by an experiment, to be carried out by his student Stafford. Unfortunately, this time the mentee does not live up to his mentor's expectations. When the experiment, which leads Cantor to the Nobel Prize, is at stake, he sets his priorities: he wants to impress his own mentor, Kurt Krauss, a Harvard professor, who suggests the candidates for this year's Nobel. He cannot risk a failing lab project and decides to repeat the experiment himself. Cantor distrusts Stafford and withdraws from him completely. Trust as the major bond between mentor and mentee seems to be only important as long as the mentee's contribution adds to the mentor's reputation.

As Stafford is dependent on Cantor, so is Cantor on Krauss. Although both are professors, their seemingly collegial relationship turns out to be pure power play. Djerassi deconstructs trust as a relational category in science since it is mainly used instrumentally. A more suitable term would be interdependence, since power in science is described as a dynamic concept: once Stafford is awarded the Nobel Prize together with Cantor, Cantor needs his co-operation, because they are expected to present themselves as a team – and Krauss eventually hopes for Cantor's support as he is yearning for the Nobel Prize himself. In Djerassi's analysis of the culture of science, relationships are more important for the researcher's success than scientific findings, but the male scientists portrayed seem mostly unable to establish and nurture relationships with their peers. Egocentricity and ambition are the main driving forces towards success – the relationships within the 'old boys' network' last only as long as its members are useful to each other in their professional careers. Outside of the academic context, the male scientists are described as lonesome and oftentimes unable to organize their everyday lives.

Even though Djerassi's culture of science is not exclusively male – he does present some female pioneers – there is no doubt that in science women are still an exception and that they have to operate outside of the established male networks. The representatives of the humanities on the other hand, whose critical approaches to science open up important perspectives for a development of the scientific discourse, are exclusively female. Djerassi's portrait of the scientific culture raises the question how science came to be a male discipline.

In her fascinating book *Has Feminism changed Science?*, Londa Schiebinger focuses on the gendering process of science from a historical perspective. She quotes research carried out by Margaret Mead and her colleague Rhonda Métraux in 1957, where they interviewed American high school students about their expectations concerning scientists. Students expected a scientist

> to be 'a man who wears a white coat and works in a laboratory. He is elderly and middle aged and wears glasses...he may wear a beard...he may be unshaven and unkempt. He may be stooped and tired. He is surrounded by equipment: test tubes, Bunsen burners, flasks and bottles, a jungle gym of blown glass tubes and weird machines with dials.' (Schiebinger 72)

The students also considered such a scientist a quasi-genius, probably neglecting his body for his mind, and most likely also his wife and children – if he has any at all. (see Schiebinger 72) This image of the scientist as a male, confused genius survived well into the 1980s as a collection of 165 drawings by secondary school-children proves. One of them is included in Schiebinger's book: it shows a middle-aged bald man wearing glasses, standing next to a Bunsen burner and thinking about a chemical formula. (see Schiebinger 74) Most of Djerassi's male scientists fit this description perfectly: They work in an old boys' network and are busy reproducing it.

Schiebinger tries to trace the exclusion of women from the scientific culture historically, looking at the development of the relationship between gender and science. Starting in the 17th, and even more so in the 18th, century, a division between the public sphere of government and the professions on the one hand and the private sphere of hearth and home on the other took place. Schiebinger calls this a "privatization of the family" and "professionalization of science." Science, which at that point moved more and more from the 'private' into the public sphere of universities, became a men's world, and "came to be seen as decidedly masculine." (Schiebinger 69f.)

Schiebinger is aware of the problem that a focus on gender aspects when looking at the history of science can make the category of gender more dominant than it might have been – this might reinforce gender stereotyping. At the same time she points out the risks of ignoring gender differences as this leaves invisible power hierarchies in place. In her view, "masculinity and femininity" do not have universal meanings beyond historical contexts and may mean many different things at different times and in different places. Class might be an issue, the national context might be as well. But

as decades of scholarship have demonstrated, however, sexual differences define powerful fault lines in our culture. ... Gender in the style of science is significant because women's long legal exclusion from scientific institutions was buttressed by an elaborate coding of behaviors and activities as appropriately masculine or feminine. Unearthing assumptions surrounding gender in science helps unearth unspoken notions about who is a scientist and what science is all about and how these notions have historically clashed with expectations about women. (Schiebinger 69)

This differentiation of the culture of science on the one hand and the culture of femininity on the other was also supported by the writings of the "complementarians," among them, according to Schiebinger, Hegel himself:

The theory of *sexual complementarity* ... fit neatly into dominant strands of liberal democratic thought, making inequalities seem natural while satisfying the need of European society for a continued sexual division of labor. ... The private, caring woman emerged as a foil to the public, rational man. (Schiebinger 70)

Women were historically not only excluded from science – as from the university in general – but also perceived as incapable of doing scientific research. According to Schiebinger, this lasted well into the 20th century. Nowadays women are part of the culture of science. There are not many of them – in the league of professors and as members of scientific circles such as national academies, but they are at least allowed to participate.

Schiebinger's findings put Djerassi's literary presentation of a primarily male culture of science into historical perspective. What he describes has a long tradition and has successfully influenced men's and women's perception of how a scientist has to be. But does Djerassi in some way write "against" this cultural imprint? Do his female scientists network differently – or even practice a female style of science? Does he create a female alternative to a male dominated culture?

In *Cantor's Dilemma*, there is Professor Jean Ardley and her student mentee Celestine Price. Jean has made concessions to the rules of the masculine culture, tying her tubes and changing her name:

I finally decided that, given my professional ambitions, I just couldn't do justice to motherhood. ... I'd say that in chemistry, or for that matter in most laboratory sciences, you just can't be a mother and get tenure during the six years you've got as an assistant professor. At least not in the big research universities. My male peers put in at least eighty hours a week. That's why many of their marriages don't work out – if they marry at all. (Djerassi 1989, 44)

Moreover, Jean changed her name from Yardly to Ardley, because she thought that this might guarantee her being named first in collective publications. Cantor had solved this problem his own way: he never publishes with anybody whose name starts with an A or B. In the end Jean adopted her female mentor's practice in putting her name last, because she came to see that being first is not the crucial factor in scientific competition. Jean's professional goal is to become a member of the National Academy, as there is only one woman in the chemistry section. She realizes early on in her career that she can only get to the top when playing by men's rules. But she has also decided that there are ways to interpret these rules and norms differently when it comes to working with her own students. Jean and Celly cooperate collaboratively; she supports Celly in the lab and they learn from each other. With Jean and Celly Djerassi suggests an alternative model of the mentor-mentee relationship where both parties bring something to the relationship which makes them a better team. Hierarchies are still there and power is an issue, but Jean openly talks about it and discusses her position with Celly.

Another successful female representative in Djerassi's fiction is Charlea Conway in *The Bourbaki Gambit,* a professor of mathematics at the University of Chicago, and a lesbian. She has also decided against a combination of profession and family, because she wanted to follow her main passion in life, which is research:

'I have no children. And no husband, either. I never wanted them. But if I'd had to choose between science and family, I know which would have won out.' 'But isn't that quite of a sacrifice?' persisted our hostess. 'One only women have to make?' Charlea shrugged. 'It wasn't a sacrifice for me. Scientific research is addictive in the most positive sense of the word.' (Djerassi 1994, 74f.)

Charlea challenges Max, the retired professor from Princeton University, starting with their first meeting, ignoring his title and addressing him by his first name without prior warning. She completely fails Max's expectations, both in terms of politeness and appearance. She dresses "carelessly" and wears "absolutely no makeup; her hair seemed to have been cut by a man's barber." (Djerassi 1994, 55f.) Charlea fits the picture Londa Schiebinger draws of successful women in science in the first half of the 20th century, where neglecting one's appearance was also a way of not being perceived as a sexual being – "the highest compliment to a woman of science was to be made an honorary man." (Schiebinger 77) With Charlea, Djerassi presents a female character who challenges traditional expectations of women's roles as mothers and style queens. Even though she revolts against male expectations of womanhood, she has accepted and adapted to the cultural norms and values of her male fellows. In terms of gender sensitivity, Djerassi presents her as rather naïve, since she is convinced that gender aspects play no role in science and that women can make it in just about any field – if they are competent enough. The Bourbaki project teaches her otherwise. (see Djerassi 1994, 74) Diana, the central female figure in *The Bourbaki Gambit,* a historian, creates gender awareness not only in her, but in the men involved as well.

Besides Jean and Charlea, who have both established themselves within the academic field, Djerassi also presents two young female scientists at the beginning of their careers: Celestine Price and Jocelyn Powers. Renu, the main female protagonist in *NO*, starts out as a promising scientist, but leaves academia in order to found a biotech enterprise, mainly because she realizes the male boundaries in the field. Jocelyn, a young student and Diana's granddaughter, becomes the lab assistant of the Skordylis group, as the private circle of scientists in *The Bourbaki Gambit* is called. Because she knows their secret, they depend more on her loyalty than she does on theirs. Although the other members of the group treat her as an equal, Max Weiss sticks to the professor-student relationship. But Jocelyn makes it quite clear to him that she is aware of her role in the power game – and that one could also interpret it differently. When the Skordylis group splits up because they can no longer agree on the rule of joint publication under a false name, Jocelyn becomes independent: she sends the two papers the group had finalized on PCR (Polymerase-Chain-Reaction) off to *Nature* without informing the group. The articles are published under her and Diana Skordylis' name. The research idea becomes famous and is even awarded a prestigious prize. Jocelyn questions the masculine culture of science – partly because she has reflected some of the norms and partly because of a risk-taking naiveté. As an assistant she has not yet made a clear decision about her future career, whether to go into research or not. Her escape to Vienna, where she becomes engaged to Sepp's stepson, also seems to be a reaction to her "intercultural" experience with the culture of science. With Jocelyn – and actually also Stafford – Djerassi suggests that for some young scientists the culture of science might be too much of a challenge – more preparation in terms of cultural awareness would be important to stay and survive in the field.

The other female scientist with a promising career is Celestine Price. She follows her high school teacher's advice to go to a big research university in order to get plugged into the old boys' network. She chooses her own way to do this through a relationship with her professor, Lufkin, who ends the relationship when he becomes scared of his feelings. But he still remains a valuable resource for her about the cultural tribe. Different from her high school mentor, he advises Celly to do graduate work with a woman, Jean Ardley: "'You might as well get a female role model in graduate school and find out how she did it. What the costs are. How her male colleagues treat her.'" (Djerassi 1989, 23f.) She follows his advice. After her graduate degree she makes another unusual decision, when she turns down an offer from Harvard. Her partner Stafford, who, after his breakup with Cantor, had accepted a position there, does not understand her reasoning. She explains her decision:

'You've been here only a few months, but you're already like all Harvard men. They think it is the ultimate place; they only have to whistle and you come.' 'But, Celly, this is Harvard. The best in the country.' 'For whom? ... Let me tell you why I say that. ... it was obvious to me that, even though they were excited about my work and its possible practical applications, they'd never have considered me at this stage – not yet a Ph.D. and no postdoc experience – if I were not a woman.' (Djerassi 1989, 166)

In contrast to Charlea, Celly is very gender conscious and has realized that an offer made exclusively on gender terms eventually will turn into a dead end street. She chooses the longer perspective of a scientist over a short-term promotion for prestige. That Celly has followed her career objectives becomes obvious in Djerassi's last "science-in-fiction" novel *NO*, where she reappears as a member of the same Scientific Advisory Board as Cantor (who chairs it). Celly is now a professor at Caltech with a small research group working on promising projects. She is selected because of her work and not because of her sex. She is the only successful woman in academia who combines children and career. This becomes possible with the support of her husband Stafford, who still goes to medical school. Whenever she is gone, he takes care of their son. How she has managed to combine motherhood and research is not made explicit – although it would be helpful for young female readers of Djerassi's novels to know. Although Celly challenges Cantor as a traditional representative of the masculine culture of science, they also have much in common: Celly admits to being very ambitious and to being infected by "Nobel lust" (Djerassi 1998, 247) as Cantor calls it.

Overall, the successful women in Djerassi's literary world of science still function along the same norms and values as their male colleagues. They are equally ambitious, hard working, more or less competitive and in search of recognition. What is different is the degree of their egocentrism and their collaborative approach when working together. They show more awareness of gender and power issues, which makes them stronger when fighting for their own position – as can be seen in Celly's case. But in terms of an alternative style of doing science, Djerassi remains as realistic as Schiebinger. "Curiosity," "ambition," "egocentrism" are just as much the driving forces for his female scientists as they are for the men in the same fields. Schiebinger disillusions her readers in regard to that question as well: "The hypothesis that women will do science differently remains just that – a hypothesis in need of testing." (Schiebinger 11) There is no 'feminist' or 'female' style ready to be plugged in at the laboratory bench or the clinical bedside. Feminist goals in science will not be realized through the invocation of cliché-ridden principles drawn from a mythical 'lost feminine.' ... It is time ... to incorporate a critical awareness of gender into the basic training of young scientists and the workaday world of science. (Schiebinger 8)

As Djerassi suggests in his fiction, this process of creating a cultural awareness can be best achieved when scientists go beyond the boundaries of their own discipline. And it is no surprise that the driving forces for such a reflective process are – with the exception of Lufkin, the only 'loser' in the academic environment – mainly women, either professionals, who are working outside of academia, or students and professors of the humanities.

In *Cantor's Dilemma*, Leah presents Bachtin's theory on dialogism to Celly, Stafford and Jean. She demonstrates that an analysis of the scientific discourse can help to detect hidden power relations in academia. It is her input that makes Stafford realize that power is an issue in his relationship to his mentor Cantor. The discussions with Leah are critical for Celly's development of gender awareness for her field.

In *The Bourbaki Gambit*, Djerassi even goes a step further. He not only shows the relevance of the research findings of the humanities for science, but creates a project where different disciplines work together. The driving force behind this idea is Diana, a historian researching the French salon which served as a framework for scientific research outside of the universities in the 17th and 18th centuries. These salons were organized by influential women. For Diana, the Bourbaki project is a way of transferring the idea of the salon into the present. She considers this applied research. She wants to test whether the boundaries between science and the humanities could be overcome:

'Don't worry,' she said. 'What I want won't interfere with your research. It will just make it more elegant.' 'And what is it you want?' I asked. Somehow, I felt that precision was important at this point, as if we were discussing the fine print in a legal contract. 'To reintegrate men and women in science; to reconcile science and the humanities, show them for what they really are: integral parts of something larger – something we hardly have a name for.' (Djerassi 1994, 66)

One is reminded of C.P. Snow's influential "The Two Cultures" (1959), where he argues that scientists and "literary intellectuals" (see Gehrke 133) need to work together in order to cope with the challenges of our society. With the Bourbaki idea Djerassi creates such an 'intercultural project' between the humanities and the natural sciences. And it is usually the humanities – represented by women – who initiate a critical reflection of the cultural norms of science. Without the women, no development would take place in and with men. Djerassi does not necessarily present a re-gendered version of science, but in his fiction it is definitely the women who raise the awareness for gender in science. In general, the female characters in his works appear to be more interesting than men. This is true for his novels; but also for his plays, and definitely for his book *Four Jews on Parnassus*. The tribute women would deserve in the scientific and academic communities is paid – to some of them – in Carl Djerassi's literature. As he states in an interview: "It's not really part of a plan, but somehow it's always the women who turn out to be the real heroes of my novels." (Nolte 1990, quoted from Gehrke 145) Somehow the process of creating women in science-in-literature has turned out to be a personal "re-gendering" project for Carl Djerassi – which might well qualify him for re-birth as a woman in his next life.

Works Cited

Djerassi, Carl. 1989. *Cantor's Dilemma*. New York: Doubleday.
Djerassi, Carl. 1994. *The Bourbaki Gambit*. Athens, Georgia: The U of Georgia P.
Djerassi, Carl. 1998. *NO*. Athens, GA: The U of Georgia P.
Gehrke, Ingrid. 2008. *Der intellektuelle Polygamist. Carl Djerassis Grenzgänge in Autobiographie, Roman und Drama*. Münster: LIT.
Gomboz, Ingrid. 1999. "Carl Djerassi – Vater der Pille: 'Im nächsten Leben wäre ich lieber eine Frau'." *15 5 +80 – Zeitung für Absolventinnen und Absolventen der Karl-Franzens-Universität Graz* 1/99: 9. Translation from this text is by author Ingrid Geluke (née Gomboz).
Schiebinger, Londa. 1999. *Has Feminism Changed Science?* Cambridge: Harvard UP.

The Academic Novel: A Personal Typology

Pierre Laszlo

The academic novel occupies a niche in the big house of fiction. Its dual function is to reveal academic life to the general public in an amusing way and to present a mirror to its protagonists. In so doing, it often aims no higher than entertainment – actually a challenging goal. My credentials for attempting such a study are a lifelong membership in academia, including visiting professorships at a dozen or so American research universities, insider knowledge from having belonged to or visited both scientific and literary departments, plus having been an avid reader of books on college and university life for a long time.

In addition to the obvious questions of whether a particular campus novel gives the impression of 'being close to reality,' and whether it is indeed funny and a good read, other issues are raised and will be addressed in this essay. Another question is why campus novels, with notable exceptions such as those by Carl Djerassi, are written almost exclusively by literary dons, with a remarkable absence of their colleagues from the sciences. Only the briefest of answers will suffice: the former, but not the latter are trained to write good prose.

"Is academic life intrinsically comical?" is a question that is fundamental to an understanding of the genre. I will point out the abundance of paradoxes in the diverse activities of college teachers. The master paradox has to do with transmission of knowledge. Teaching can be understood and is understood as entailing two diametrically opposed activities: (i) conveying already existing knowledge and pouring it into the heads of students, an inherently conservative activity symbolized by university libraries; and (ii) pushing the frontiers of existing knowledge and contributing to the advancement of science, an inherently subversive activity. But there are many other paradoxes of academic life, such as (to put it in capsule form) extremely smart people behaving like idiots – a recurring feature and source of mirth in campus novels.

Why did campus novels flourish during the second half of the twentieth century? Why is it a predominantly Anglo-Saxon genre? The answer, I submit, lies in the rise of research universities in the post-World War II period. In 1945, Vannevar Bush presented his report to the President of the United States, *Science, the Endless Frontier*. As a consequence, the National Science Foundation was set up in 1950. At the same time, the National Institutes of Health, that date back to 1930, were much expanded. The system of research proposals and grants, which has endured and thrived since, was put into place. The launching of the Sputnik by the USSR in 1957 stung the American collective psyche, and it responded with an increased emphasis on both science education and production. The existing American colleges had been modeled on British counterparts such as Oxford and Cambridge since the eighteenth century, inclusive of their architecture. Thus, the expanding American research universities continued to use the same mould which produced a tangible tension between traditional liberal arts colleges and the newer institutions. Entrepreneurship in the latter conflicted with scholarship in the former, which provided academic novels with a rich satirical vein, whether one made fun of erudite dons totally lacking the ability to face the practicalities of modern life, or conversely of operator types with a blind spot to any historical dimension.

Anxieties

The potential disasters in an academic biography are so numerous that it seems best to list just a few without comment in tabular format:
    A losing a grant
    B fraud by a coworker
    C losing one's lecture notes
    D losing one's slides
    E losing one's computer
    F losing the manuscript of a book
    G a hard drive gives up on you
    H getting a bad review of one's book
    I inability to answer a question at a seminar
    J not being granted tenure
    K being scooped
    L losing one's good name
    M quoting inaccurately/being quoted inaccurately
    N losing a few square feet of laboratory/office space
    O failing to be appointed to a major committee
    P inability to attend a conference due to illness
    Q misplacing library books
    R missing a deadline
    S being vilified in print & inability to answer
    T forgetting one's class
    U inability to locate an unfamiliar classroom

V in debate, being slow on one's feet
W failure to find irrefutable arguments
X going through a dry spell
Y losing publicly one's temper
Z turning into an alcoholic

This is the substance of nightmares. Academic novelists tap into these familiar threads for their plots and for an easy source of derision.

Ethnography

One might argue that the academic novel resembles ethnographic writings. Its standard categories are the natural milieu and the habitat, the daily life, social organization, religion and beliefs, rites of passage, dress, language and culture, and sex.

A quick survey of this approach, as it applies to academia, will show the rich material writers of campus novels are able to draw from. The natural milieu and the habitat of the tribe are the campus, i.e., a group of usually stone buildings, often in Gothic style, set on large lawns in or at the periphery of a relatively small city. The Oxford and Cambridge colleges were the archetypal models in the architecture of colleges and universities elsewhere, especially in the United States.

The social organization is a rigid caste system: the professors are at the top of the pecking order. Below them are junior faculty, readers, lecturers in the British system, associate and assistant professors in the American system. The third tier is that of the graduate students and postdoctoral fellows. On the same level, more or less, stand undergraduate students whose education is the ostensible goal of the institution.

Daily life is run according to equally rigid rules and depends on the group considered. In the American system, a professor spends his time giving classes, writing papers and research grant proposals, preparing for trips devoted to lecturing and consulting, attending meetings – all of which makes for an exacting schedule. At the end of the day, a small glass (or two) of an apéritif provides the needed solace.

In contradistinction, a graduate student will be spending a major part of his time at the computer, will be involved in looking for his next job, will be doing some laboratory work, will also perform some teaching to undergraduates. This is also an exacting schedule and, at the end of the day, one (or two) drinks may equally be called for.

Religion and beliefs are centered on the written transmission of knowledge: "publish or perish" is a strongly-held creed, together with a whole set of other strict rules such as "do not publish identical material twice," "do not plagiarize," "cite earlier work." Academics even carry heavy intellectual loads around; for example French Theory, associated with eminences such as Jacques Lacan, Roland Barthes, Jacques Derrida, Michel Foucault or Jean Baudrillard.

Language is also remarkable in its importance. While English is the lingua franca, especially in writing, the spoken language has a domineering place in campus life. Because of the necessities of lecturing, many dons show impressive command of it. Some revel in wordplay and turn public places such as cafeterias into locales for punning displays. Faculty meetings on every level provide the occasion to further hone one's argumentative skills.

Academic culture is another attribute worthy of the anthropologist. General culture originates (in undergraduate student life) in the Oxbridge colleges. Faculty members make ostentatious displays of their general culture at socially important dining events, the archetype for which is conversation at the Oxbridge high table. Moreover, each academic discipline has its own, specialized culture which is constantly rehearsed and played out at conferences, consultancies and even in law court, in the form of expert testimonies. Academic culture is thus also expressed in priority contests and, to pre-empt them, constant salesmanship and one-upmanship.

Dress, for undergraduate students, is not a uniform to be found only on campus. They constitute a subset of youth, as a culture, and thus tend to follow its very strict and quickly evolving codes. Teachers, conversely, have their own cherished mannerisms. British tweeds still endure. There is also the occasional eccentric costume – the campus is a privileged shelter for eccentrics of every variety, male or female. Ceremonial gowns, worn on special occasions, are a pseudo-clerical form of dress, referring to the religious origins of colleges and universities.

Romantic involvements occur among both groups, the faculty and the student body, in both the homosexual and the heterosexual mode. There is a taboo, however, against a professor romancing an undergraduate student and thus abusing his or her elevated position within the university.

The comical mode

What is obvious raises the question of the reason for the obviousness. Why is campus life inherently funny rather than, say, depressing? Why does it lend itself so nicely to being written about in novels? Is the literary form of the novel somehow congenial to affectionate descriptions of life on a campus?

The novel, because it is a narrative and a narrative appealing to the reader in his or her imagination to view it as 'real,' and because it has the ability to describe behaviors of all kinds, including surprising acts, can elicit a whole register of emotions. One can be moved to tears. Or feel a lack of sympathy for the characters. Laughter is one such emotion, which most academic novels employ.

The sheer improbability of a reported gesture, expressed in dialog or in an episode, may induce the delight of surprise. If furthermore it is out of character, this may trigger an ecstasy of laughter. Professors can be eccentric enough, in actual academic life, as to make them comical. The novelist has only to act as a reporter of their actual deeds, going through some trouble to conceal their real-life identity, prior to staging their antics. To the reader, departure from an established behavioral norm can be comical. Consider, for example, the absent-mindedness of a scholar possessed by an idea which has captured him, which he seeks to make better sense of. That can be extremely funny. There are plenty of stories of this type, for instance Einstein falling in a ditch dug by road workers upon walking home in Princeton or Norbert Wiener, about to lecture to his students at MIT, pushing aside a person blocking his way to the blackboard – in fact another lecturer, Wiener having entered the wrong room.

Thus, there exists a whole oral tradition of such stories. Academics tell them gleefully to one another. The late Vlado Prelog, a Nobel prize winner in chemistry, was renowned for his repertory of stories and jokes. Indeed, Carl Djerassi, whose skills as a novelist will be commented upon below, is himself a very gifted storyteller. He knows that one of the secrets is not to get quickly to the point, but to delay such satisfaction on the listener's part and drag out the story quite a bit in order to prepare for the unexpected punch line or ending. Thus, I cherish a memory of Carl's mirthful narrative of how, after the 1987 Nobel prizes in chemistry had been selected in Stockholm, a Swedish colleague called Donald Cram in Los Angeles to announce that he was its winner – only to meet with continued hilarity of the person on the phone: he had reached a homonymic Donald Cram, dentist and not a chemist.

From the grapevine to the inside of the scientific community: it bonds to some extent through dissent. Argument and debate are essential activities characterizing the group. Violent controversies between scholars are not infrequent. Their colleagues take stands, more often watch avidly the trade of barbs and even insults. It is definitely a spectator sport. To be on the sidelines is easily laughter-inducing. Indeed, score may be kept over how frequently one ridicules the opponent. Such violent disputes by academics on litigious points go way above the heads of outsiders, journalists in particular, who lack the background for understanding what is at stake. Hence, such behavior is totally exotic to people outside the field. They are the very stuff of low comedy. No wonder academic novels report such quarrels, in either incomprehensible genuine form, or caricatured into shouting matches of mutual abuse.

Freedom

By and large, it is academics who write academic novels. This statement raises the more interesting question of why some academics "waste" their time in such a manner. My contention is that it provides them with considerable freedom of manoeuvring in which they are able to air their personal views, not only to their fellow-academics, but also to the world-at-large as represented by the ordinary reader.

But why vie for such oxygen? Because the academic life, rosy though it may seem from the outside, is tough. Classes have to be met, day in day out. Even though you were attending a conference in Manitoba, you have to be back the following morning in your classroom in the Midwest or in Western Europe. You are not allowed to be sick, lectures have to be given, engagements have to be met. To find some time for work is a constant challenge between committee meetings, students coming in to protest a grade they might have richly deserved, telephone calls, a stream of ingoing and outgoing e-mails and reading professional journals.

The life of the scholar has its hardships. The profession demands publication of books. The work required is inherently destructive to family life: this assertion is borne out by the uniform acknowledgment in print of the sacrifices imposed upon one's spouse. Then, there are all the constraints, gradually learned and internalized. The writing of a specialized publication has to follow strict guidelines. The presentation of a paper, as a lecture or at a conference, is also a strictly regulated genre. Even the more informal moments are, in actual fact, very much codified. For instance, participation in a faculty lunch of any sort demands witty interventions, consisting of a combination of alertness, erudition and punning. This requires training.

If publishing one's work is fraught with difficulties, what about lecturing to students? Each class is framed by a rigid time slot. It needs extensive preparation, from one to ten hours (or more) per contact hour. A hand-out has to be prepared for the benefit of the audience. While the contents have to be planned in minute detail, the actual delivery not only has to allow interruption from the students, it has to elicit their input in order to enlist their active contribution.

Writing a novel is truly one of the few ways an academic can get a little relief from such a hard life. He or she can transmute an experience into a moral lesson of sorts, transpose witnessed or undergone indignities into satire, and demote all-too-real episodes one has evinced into healthy laughter.

Typology

However light-hearted, academic novels communicate a rather wide variety of serious concerns. Pride of place goes along with a sense of community, a worldwide network of scholars and scientists with shared values who embody hope for averting humankind from disaster. Fred Hoyle's *The Black Cloud* (1957) together with C. P. Snow's *The Masters* (1950) and Alison Lurie's *Love and Friendship* (1962) exemplify this strand, contemporary with the height of the Cold War, and the Cuban missile crisis. Endearing portraits of academic eccentrics are those of Nabokov's *Pnin* (1953) and of Barbara Pym's *An Academic Question* (1986) or Mary McCarthy who, in writing *The Groves of Academe* (1951), made a subtly eloquent case in praise of dignity, as both fragile and utterly precious. Closely related to academic novels depicting academia as an oasis in the wilderness are a list of books heavily self-referential, such as those authored by David Lodge – *Changing Places* (1975) is a prime example – and Malcolm Bradbury.

Female authors have used the medium to popularize feminist issues from early on, when universities were still under the governance and dominion of white males. Examples are the series of whodunits – the so-called Kate Fansler novels – written by Carolyn Heilbrun, a professor at Columbia University who used the pseudonym Amanda Cross. A very early example, also a detective novel, is Dorothy Sayers's *Gaudy Night* (1935). To that strain also belong, to quote only these two titles, Gail Godwin's *The Odd Woman* (1974) and Jane Smiley's *Moo* (1995). A distinctive sub-group conveys nostalgia for the small, liberal arts colleges which, in the age of mass education, might be receding into the past. Representative titles are Richard Russo's *Straight Man* (1997) and P. F. Kluge, *Gone Tomorrow* (2008).

and Carl Djerassi

The beginning and the end of a work of fiction is always very revealing. Take Djerassi's *Cantor's Dilemma* (1989) as an example.

[Opening paragraph] 'Damn,' he muttered as he pressed his hand against his throbbing knee. He hobbled to the bathroom, feeling his way with his right palm along the wall. It didn't take expertise in neurobiology to know that photochemical stimulation of the retina was the surest way of waking up. (Djerassi 1991, 3)

[ending:] P.S. On rereading this letter, I just realized that I neglected to write about the confirmation of *your* experiment. Given the faint cloud hovering over Stafford (which, incidentally, you could dispel instantly), I suggest that we let that one dangle as well. there really isn't any hurry – after all, you, like Stafford, have already reported your work in a Nobel lecture. (Djerassi 1991, 227)

One can make a case that these two small segments neatly frame the whole plot, together with its roots in Carl Djerassi's personal life and character. The very first sentence is autobiographical; the author indeed suffers from a knee injury incurred in a skiing accident. Such a disability connects with, at the end of the book, in the postcript (in cauda venenum!) to a letter sent by a colleague at Harvard to the narrator, the allusion to the flawed experiment which earned him, jointly with his postdoctoral associate Stafford, a Nobel Prize. This experiment is non-reproducible. It is possible that Dr. Stafford made it up and that it is thus invalid.

The opening paragraph is the account of an awakening. A voice, anonymous at that point, swears. A body makes its way to the bathroom, presumably to relieve himself. We identify him, in the third sentence, as a scientist, speaking to himself, now in omniscient narrative voice.

Another beginning, that of *The Bourbaki Gambit* (1994), returns to the opening of closed eyes. Lucidity is an endearing attitude characteristic of Djerassi's 'being-in-the-world.' This opening paragraph starts with the quote of a joke, "'A visionary is a person capable of seeing with closed eyes.'" (Djerassi 1996a, 1) Again, the narrator – yet to be identified – has his eyes closed. These opening sentences belong to the stream of consciousness type, in the tradition of James Joyce. The same book closes on an image which one might fancy to be out of one of those seventeenth century Dutch still-lives, a loving description of the difference in texture between butter and margarine.

In *Marx Deceased* (1996), the author again uses devices of this tye. He starts with a quote, followed by what amounts to a stage direction ("the voice on the telephone was brusque"), which Carl's friends cannot mistake as at least slightly autobiographical, and closing with an image ("the R rolling like a Ferrari in first gear") in *Time* magazine-style. (Djerassi 1996b, 3)

In the ending, the penultimate sentence is a thought, attributed by the omniscient narrator to an investigative reporter (for the *Village Voice*): "she recalled that Janus was also the Roman god of all beginnings, and she wasn't ready to start all over again" (Djeraasi 1996b, 218) – which hints at a cyclical time, linking beginnings and endings, which is exactly what the current reading is about.

Carl Djerassi has a very good feeling for dialogue. The novelist can be a chronicler of his age. Due to his Bulgarian background, his Viennese childhood and his American formative years, not to mention his stay in Mexico City at the beginning of his career, he became familiar with several languages. These experiences left him with a life-long curiosity for languages. Hence, his novels can be seen as documentaries for the way in which scientists express themselves orally. They record speech in its idiosyncrasies and in its markings by a specific time: how some ways of speech, some expressions, are typical of a given place and period. Djerassi has set them down, before they vanish, because they reflect a culture of a given time period, namely American academia during the second half of the twentieth century. One of the finest examples appears in *The Bourbaki Gambit*, in an exchange between a professor and his dean:

'Unfair, Diana. All you do is read and write.' / 'All?' / 'I don't mean to say that's nothing,' I added quickly, retreating from that spoon. 'But it does not require much in the way of equipment. Beyond the library.' / 'And computers.' / 'All right, computers. Of a sort. But what are you driving at?' / 'I'm trying to get some idea of the budget you'll need. Beyond personnel. I assume there will be laboratories, equipment.' / 'Actually,' I said 'There won't.' / 'No labs?'/ (38)

There emanates from academic novels an overall feeling of loss and expulsion from the good life, of a lost paradise due to the invasion of the vulgar and philistine. A joke that went round campuses embodies it nicely: God and Satan are meeting. They are idle. Thus Satan dares God to play a game with him: Can he create something so perfect that he, Satan, would not be able to spoil it? God answers that the Devil ought to know better than play games with Him, the Omnipotent. The Devil just tells him to go ahead and make His move. God thus is compelled to create the most perfect object he can think of, the university professor. The Devil, smiling contentedly, answers with an all-too-easy countermove. He brings to life in turn the Dear Colleague.

Works Cited

Bevan, David G. 1985. "Images of Our Tottering Tower: The Academic Novel as a Metaphor for Our Times." *The Dalhousie Review* 65/1: 101-10.
Carter, Ian. 1990. *Ancient Cultures of Conceit: British University Fiction in the Post-War Years*, London: Routledge.
Djerassi, Carl. 1991. *Cantor's Dilemma.* New York: Penguin.
Djerassi, Carl. 1996a. *The Bourbaki Gambit.* New York: Penguin.
Djerassi, Carl. 1996b. *Marx Deceased.* Athens, GA: U of Georgia P.
Moseley, Merritt. 2007. *The Academic Novel: New and Classic Essays.* Chester: Chester Academic Press.
Showalter, Elaine. 2005. *Faculty Towers: The Academic Novel and Its Discontents (Personal Takes).* Philadelphia: U of Pennsylvania P.
Womack, Kenneth. 2002. *Postwar Academic Fiction: Satire, Ethics, Community.* Basingstoke & New York: Palgrave.

Science-in-Poetry

Private Words Addressed in Public: Carl Djerassi's Poetry at the Science-to-Humanities Interface

Federico Maria Rubino

Carl Djerassi started his by now burgeoning literary career by writing and publishing short 'self-therapeutic' poems which were later collected into a small book, *The Clock Running Backward*s (1991). In so doing, he resembles one of his fictional characters, retired Japanese biochemist Hiroshi Nishimura, who eventually finds new a intellectual life as a writer of *haikus*. (Djerassi 1994, 61)

The 22 poems in *The Clock Running Backward*s are divided into three sections: an autobiographical group (1-11); an 'Aesopian' group (12-17); and one expressing Djerassi's appreciation of the the American poet Wallace Stevens, in whom his late wife, Diane Middlebrook, specialized, and of the German-Swiss painter Paul Klee, of whose artwork he is a prominent collector (18-22). Thus Djerassi's poems cover a number of topics, both autobiographical and 'public,' but as poetry is private, it has raised so far comparatively less interest than his achievements in Science-in-Fiction and Science-in-Theatre. However, just as Djerassi's young Nobel Laureate Stafford quotes T.S. Eliot in his Nobel banquet toast,

'These are private words addressed to you in public.' (Djerassi 1989, 188)

similarly Djerassi's poems, introspective as they are, are possibly much less private than a first reading might suggest. Therefore, while working on the Italian translation of Djerassi's novel *Cantor's Dilemma* and of the poems in *The Clock Running Backwards*, I tried understand his appeal to the broader audience beyond the Djerassi *aficionados*, readers interested in science-in-literature and literary scholars. In fact:

For voyeurs,
For seekers of private windows
Into a poet's mind
Concordances are custom made.
(Djerassi 1991, 48; 1-4)

In this contribution, I will describe points of contact between Djerassi's works, especially in poetry, and some of the 'hard' issues in the relationship modern professional science entertains with the broader world. My perspective is that of a trained chemist working in the Italian academic food-chain, in a field where Djerassi was, many decades ago, a groundbreaking master.

I will first focus on both the power and the weakness of scientific metaphors which occur not only in literature but are also commonplace in everyday communication. My second concern is the socialization of scientific and technical professionals who face a transition from the status of an intellectual élite to that of an ordinary, although skilled, workforce, increasingly devoid of decision-making power. Finally, I will reflect the similarities and differences between scientific and 'artistic' creativity and the meaning of this difference for human ambition, pride and the yearning for professional recognition.

In the following two telling quotations, there is a strong parallel in the use of central chemical metaphors. In the opening poem of Djerassi's collection, "Exhortation," he likens the poetic process to procedures in chemistry:

Synthesize a poem?
Distill its essence?
Filter the impurities?
Evaporate it to dryness?
(Djerassi 1991, 9; 3-6)

On a gentler, very personal side, most natural scientists would take as the ultimate token of mutual affection Djerassi's funny, tender poem title: "'You wash this shirt like a chemist.'" (Djerassi 1991, 26) Another professional chemist-turned-writer/poet/playwright, Primo Levi, similarly reflects the function of metaphors of chemical origin in the commonly used language:

Anche il profano sa che cosa vuol dire filtrare, cristallizzare, distillare, ma lo sa di seconda mano: non ne conosce la "passione impressa", ignora le emozioni che a questi gesti sono legate, non ne ha percepito l'ombra simbolica. (Levi 12)

Even the layman knows the meaning of 'filtering', 'crystallizing' or 'distilling', yet his knowledge is second-hand, deprived of its 'embedded passion', devoid of the involved emotion, of its symbolic shade.

Born out of basic human needs, chemistry, since its alchemic beginnings, used words and concepts rooted in the common world of human practical activities and turned them into the description of its efforts to separate different kinds of matter aiming at understanding their differences and similarities and at exploiting their different properties through operations such as evaporating, filtering, crystallizing, later distilling and finally synthesizing (*i.e.*, making desired molecules through a rationally designed course of action). However, metals rust, organic matter ends in degradation through rotting and fermentation, and human operation on matter may yield undesired results, such as our favourite sweater's felting or Levi's unwanted polymerization.

To probe the composition and the behaviour of matter, chemists use physical ("spectroscopic") methods to 'interrogate' their specimens. Djerassi pioneered the the strategies to combine the results obtained from the individual analytical techniques (cf. Djerassi 1990; Williams & Fleming 1995); a task which needs 'spectroscopic introspection.'

Similar operations and a similar language also apply to events of human life and to psychological processes, thus creating a rich field for the metaphoric use of concepts and words of chemical origin to describe inner human behaviour. To reach happiness we 'filter' bad experiences and undesired thoughts 'evaporate' (or at least we expect them to do so); a world of life experiences 'crystallizes' into strongly held habits, and we 'distil' our wisdom to pass it on to our children. The ability of chemical entities to interact with others to yield specific products and physical phenomena has long beeen employed in literature to gain insight into human behaviour, from Goethe's *Die Wahlverwandtschaften* to Edgard Lee Master's *Trainor, the Druggist*.

Another chemist-turned-poet, Roald Hoffmann, stretched the analogy of random intermolecular interactions to those of humans in several beautiful poems, concluding that

Men (and women) are not
as different from molecules
as they think.

The soft touch a chemist needs to handle delicate glassware is the tool Djerassi uses to show physical affection ("Suck your tongue / The way I suck pipettes?"; Djerassi 1991, 26; 9-10) and love.

Such examples can also be read in view of the schizophrenic success of scientific and technological metaphors manifested nicely in very word *chemistry* itself. *Chemistry* is the artificial manipulation of supposedly *natural* entities and processes, wicked both in itself and as a companion of men's greed in the exploitation of 'innocent' Nature. This essentially chemophobic view provides neither for an understanding of the essentially *chemical* nature of the natural world nor for the appreciation of the benefits given to humankind by the beneficial applications of this knowledge of Nature, both on the material side and of that of intellectual understanding.

Carole Angier's controversial biography uncovers the metaphoric power which Primo Levi assigned to chemical concepts, starting from the subject of his

Tesi di Laurea, Walden's Inversion, to Giacomo Ponzio's polemic on the 'double oscillation' (which we now know as 'inversion of configuration'), to the 'double bond,' promoted to the symbol of 'organic life' and of living events which necessarily involve the making of 'bonds' between people as opposed to 'inorganic,' exclusively natural events. (Cf. Angier 157 and 207)

Indeed, chemistry is now the key to explain even human emotions and intellectual activities both from the scientific and from the metaphoric point of view. Steady progress in neurophysiology and the identification of the role of an increasing number of natural brain-acting substances on human behaviour has popularized a layman's image of individual and inter-personal life events as the mere consequence of *chemical* neurotransmission events, *i.e.* of *brain chemistry*. In fact Djerassi himself speaks of the significance of "chemistry" in establishing a relationship. (see Djerassi 2005) Quiet men in peaceful times are called to adrenalinic experiences, testosterone soaks Silicon Valley and the infidel spouse calls for indulgence due to congenital lack of oxytocin.

We may of course widen the reach of this perception to other words which derive from the specialized jargon of practical science and which are now commonplace metaphors, often at the edge of (or strained even beyond) their expected meaning. We may think of molecular genetics, of modern physics and cosmology. In Italy, it has become a quite common everyday language. When typing '*è nel nostro DNA*,' Google yields more than 3 million web references, spanning from political jargon to team sport and it generally means: "it is in the tradition of our group." Searching the web for corresponding examples in English, I came across the statement from U.S. House Republican Conference Chairman Jeb Hensarling (R-TX) who said it is "contrary to our DNA" to raise the debt ceiling. Professor Cantor, the chemist-turned-biologist in Djerassi's novel, is certain '*down to [his] quarks*' of the correctness of his cancer theory. (Djerassi 1989, 95)

Chemical and other scientific metaphors (such as the zoomorphic ones of Aesopian fame, most of which have faded away in the light of modern ethological studies) share with all others the origin from common-life experiences and are powerful means to express human beings' relationship to the world. Since they are also, of course, not neutral, they often have the ambiguous power to be used in multiple, including nonsensical, ways. The potential abuse of the metaphoric use of the scientific jargon to the point of intellectual dishonesty can be appreciated by the Sokal episode, in which a physicist wrote and had accepted for publication in a renowned scholarly journal of postmodern cultural studies (one Djerassi's character Leah Woodeson would select for her academic publishing) a completely fake article in which a *bona-fide* scientific lexicon and concepts were used in a voluntarily non-sensical way to highlight lack of intellectual rigor in some fields of social sciences. (Cf. Sokal affair)

Djerass's poem "Diary Entry" also serves as preface to his scientific autobiography edited by the American Chemical Society. Its main question, relating to the participants of a conference of chemists

How did they come to chemistry?
What do they do besides chemistry?
(Djerassi 1991, 14; 1-2)

is one of increasing importance when we discuss the ethical as well as professional training of the contemporary scientific community and the relationship which professional scientists entertain with the lay world. As scientific knowledge becomes more extensive, deeper, and more complex at an increasingly fast pace, the intellectual capacity of an individual to cover broader fields of understanding than one's own niche in the profession is lost ever faster and gone forever is the ability to communicate the meaning of scientific achievement not only in terms of its practical reference to welfare but also in those of its effect on professionals' feelings and behaviour. Scientists thus experience detachment from the wide world and feel a mounting public distrust – often misunderstood as anti-scientific – as soon as the subject of their research leaves the walls of the laboratory and approaches the territories of public concern. This latter is a central point in Djerassi's Science-in-Fiction and Science-in-Theatre texts: to explain the 'tribal' behaviour of professionals in contemporary science not only inwardly but also when observed in public.

'Pure scientist, you look with nice aplomb.' (Djerassi 1991, 47)

A scientific vocation has long been considered bound to the search of truth, to intellectual integrity and unselfish dedication to knowledge as well as to the public good. This is particularly apparent in the higher esteem which has long accompanied the profession of medical researchers compared to medical practitioners. The feeling of superiority among people who practice professional science along these guidelines is the "nice aplomb" in Djerassi's line. However he suggests that this is more fictional than real ("you look"), since scientists can be as greedy, as quarrelsome, as ambitious and as untrustworthy as other intellectual professionals, including poets. This is indeed a problem at a time when there is a greater need of scientific professionals not only in academia but also in the increasing number of profit-driven enterprises whose economic success is tightly bound to scientific and technological knowledge and in the corresponding bodies of experts needed in the public sector as advisors to decision makers and as controllers in an increasingly technology-driven world.

This point has been addressed by the few 'fiction' writers, often active researchers themselves, who, not unlike Djerassi, describe the life of academic scientists in the 'hard' sciences. Among literary contributions, two in particular, are Italian scientists belonging to very different generations. The late Renzo Tomatis, a founder and long-time director of the International Agency for Research on Cancer, published a journal of his experience as an *émigré* scientist in the USA in the 1960s which also dealt with the perceived differences of scientific careers in the two countries. (see Tomatis 1963) His scientific research aiming to prevent deadly human disease is challenged both by corrupt academic practices in his home country Italy and by the ruthless competition and lack of idealism in the United States, both of which he struggled to highlight to the last of his life. (see Tomatis 2011)

A few years ago, a much younger Italian university *ricercatore*, Piersandro Pallavicini, published a short novel presenting the ambiguous life of a successful Nobel Prize-level Italian chemistry professor who leaves a well-funded research line of industrial interest to dedicate himself to the search of drugs to fight AIDS in Africa. His professional choice to switch his research priorities parallels a complete reshaping of his former ruthless professional behaviour and of his affluent, sexually boundless lifestyle. (see Pallavicini) The comic strips of Jorge Cham in '*PhD piled higher and deeper*' (Cham, Coelho) show the final evolution of would-be-scientists as exploited employees in a precarious material situation whose problems are unemployment and small salaries for endless working hours rather than 'ethical' issues.

It is thus wishful thinking to expect that the bulk of the scientific workforce may contribute to narrow the 'Two-Cultures' gap: there is no criticism of 'tribal' behavior nor on the directions individual research trends take. Science as a part of an elegant *Bildung* concept as shown in Djerassi's mid-20th century professors with the sophisticated cultural background of *Mitteleuropa* has little place in those trained in the treadmill of Celly's fast lane PhD.

A rather crude example, in the *Death of a Salesman* mode, of a guiltlessly aborted scientific career, that of Dr. Douglas Prasher, was reported in the press as a heartbreaking story at the fringe of the 2008 Nobel Prize for Chemistry; it would be worthy for a scientist-writer such Djerassi to tell. The prize was awarded to Osamu Shimomura, Martin Chalfie and Roger Tsien "for the discovery and development of the green fluorescent protein, GFP." The discovery by Chalfie and Tsien was facilitated by receiving the gene for the protein as a gift from the discoverer, Dr. Douglas Prasher. The latter ran out of grants, had to shut down his lab and to leave science, ending to earn his living as a cab-driver. (see Prasher nytimes.com and Prasher discovermagazine.com)

It is in this context that the emerging pedagogical discipline of the Medical Humanities has gained increasing recognition in the medical schools, aiming at a better training of medical doctors-to-be to face bioethical issues using various forms of artistic expression: literature, theatre, cinema, visual arts. A corresponding effort towards trainees in the 'hard' sciences is, as far as I know, still largely unknown, though professionals in those disciplines will experience the burden of ethical dilemmas as compellingly as those of the medical profession:

But he,
Master of chemical mutations,
Whose alchemy touched millions,
Could not ... transform himself.
(Djerassi 1991, 55; 33-37)

Djerassi himself was always aware of the issue, to the point of offering a special course titled "Ethical discourse through science-in-fiction" to trainee scientists at Stanford, from which *A science renga* published in *Science* emerged (Djerassi 1998, 511; Aldston Jr.) As he recognizes, the 'hot' issues in scientific ethics may be very different from country to country and over time, as much as scientific fairness and freedom are challenged in entirely different ways by fast-changing economic and societal pressure.

Central to Carl Djerassi's oeuvre are, of course, the similarities and differences of two modes of creative thinking and accomplishment, those of the scientific and those of the artistic enterprise.

Poet – don't you know
There are as few scientists
With aplomb
As there are poets?
(Djerassi 1991, 47; 13-16)

Wallace Stevens' line 'Pure scientist, you look with nice aplomb' is taken by Djerassi as the title of a poem, in which he agrees to the rather obvious idea that the *prima donna* behavior often attributed to 'poets' is also likely to occur in supposedly 'pure scientists,' since it stems from human nature rather than from specific activities.

This topic, which is highlighted in Cantor's extensive use of Eliot's lines at the Nobel banquet and conference speeches, is addressed from opposite sides in *Bourbaki's Gambit* and in *Marx, Deceased*, the latter being (in my opinion) the best and most challenging of Djerassi's five novels. Whereas fiction writer Stephen Marx is able to start an entirely new life of literary production in a new place and under a new name while gleefully watching his reputation as a supposedly deceased author in the opinion of literary critics, this is hardly possible among scientific professionals due to the 'tribal' nature of the scientific profession. Essentially, scientists need to progressively build their credentials in a specific field and have them accepted by their peers who show little, if any, appreciation of outsiders, *amateurs* and mavericks. Few scientists dared to start entirely new, yet productive careers in completely unrelated fields of knowledge.

Scientific discoveries as such, even seminal ones, are not in principle connected to their specific authors. It is in fact not relevant for science practitioners, but only for science historians and epistemologists interested in who discovered what and in which priority. Once a discovery belongs to the world of scientific knowledge, it is there forever while the pride of the discoverers lasts only as long as their earthly lives – if it is even recognized within their life span.

However, to highlight that professional pride stems from a deep need of the human nature rather than from specific activities, Djerassi and Hoffmann in their play *Oxygen* have the 18th century founding fathers of the chemical science, Lavoisier, Scheele and Priestley, debate at length the priority of the discovery of oxygen and Lavoisier struggles for priority over his contemporary colleagues whom he sees as competitors, much before scientific research was established as a professional job.

In *Bourbaki's Gambit*, a novel set in the present, the struggle for individual priority of one of the elderly scientists to have his invention of the Polymerase Chain Reaction recognized is the trigger which leads to the premature dissolution of a team which had carried out its work in anonymity and collectively, attributing it to a fictitious Diana Skordylis. They thus fail to demonstrate that, although forcefully retired, they are still professionally alive and scientifically creative for the sake of knowledge and no longer for personal pride.

Bourbaki's Gambit sketches a profound difference between scientific and artistic creation and the different role and destiny of senior representatives in the two fields. The works of artists such as Chagall, Picasso, Pirandello or Hermann Hesse, who were creative until a very late age, are not endangered to fall into anonymity since artistic creation is fully connected to the creator's personality, to time, place, conditions, and gender, and there is no surrogate for their individual performances. Most science, quite to the contrary, is a collective enterprise in which the contribution of individuals is hardly recognizable, nor is this desirable. In Stafford's own words to the journalist Lundholm, "If I were the only one capable of performing that expariment, it would have no operational meaning." (Djerassi 1989, 179)

On the contrary, the destiny of accomplished and successful senior scientists is nowadays more often than not to become research managers, fund raisers, actors in science politics, governmental advisors, and *gurus* in the public arena rather than remaining active and creative in their fast-moving fields. Here, the younger researchers' faster pace makes them face obsolescence and taste the truth of the 'dwarfs over the giants' shoulders' metaphor: they are now 'the shoulders,' no more the eyes.

Yet even the most selfless scientist lives in the less-than-pure world of a structured job, in which productivity, rather than disinterested pursuit of knowledge, is the chief aim and even the glories of the past need to justify their present role. A merger of the traditional tribal structure of the academy and its current 'industrial' structure, as bluntly depicted by Professor Jean Ardley and recklessly practiced by Professor Krauss in *Cantor's Dilemma*, now allows seniors to survive the dimming of their professional leadership.

After Carl Djerassi's decision to leave cutting-edge scientific research for literature and then playwriting, his explicit early aim was to counter scientific illiteracy and mistrust through a subjectively fair description of contemporary science professionals, of their environment and of their relationship to the 'outer' world. His fiction, theatre and poetry thus communicate to readers his conception of the world's dynamic derived from life experiences encompassing most of the crucial events of western 20th century, both in the harshness of its historical development and in the enthusiasm the scientific and technological accomplishments of which his own are no little part.

But he then came to answer the question thrown onto Oppenheimer in Heinar Kipphardt's play *In the Matter of J. Robert Oppenheimer* (1964). Kipphardt's characters were world-famous physicists involved in the development of the atomic bomb, a scientific event of historical consequence and with well-understood ethical dilemmas. Djerassi's are high-ranking or budding academics, scientific and technocratic officials, whose times are those of the Golden Age of mid-20th century, explosive with science-driven technological development and midwifing the rise of an entirely new working class, that of research personnel. Their increasing number, their diversity in social conditions and cultural backgrounds and an enhanced awareness of the individual and, increasingly, collective strains of this new profession is at the core of the recent rise of a topic in genre fiction, that of lab-lit (Lab-lit; Summer reading; Rohn) which is not new but now increasing in number of titles and in variety of plots. *Nature* itself, the journal to which Cantor delivers his report of his Nobel experiment, now regularly publishes a one-page short story authored by a (usually talented) story-telling colleague. Lab-lit often questions power relationships (one of Djerassi's major topics) and other relevant issues from the perspective of insiders, such as are workers in the field of professional scientific research. A similar pathway has also been highlighted for the contemporary *medical drama*, a recent evolution of topics long covered since A. J. Cronin's *The Citadel* (1937).

Djerassi's characters and story plots support his personal optimistic view that even individual flaws do not impair the essentially healthy, competitive, trusting and all in all 'positive' nature of the scientific system. The behavior of the 'villains' (such as Krauss in *Cantor's Dilemma*) in the end damages them much more than the system. Whether one fully agrees to his conclusions depends on one's life experience, expectations and ideals. What seems to be true is that the current narrowness of scientists' *Bildung* (Stafford passed his foreign language requirement in Fortran) is of little help to open their minds and that fiction – and poetry –, employed not only as an evasion genre but also as a case study, may be of a great value in the education of scientific professionals. Also educated laypeople can expand their *Bildung* by understanding that contemporary professional workers in the scientific field are much different from the Mad Scientist or the "Telethon-style" media image.

Djerassi's discourse, based on his rather exceptional scientific and professional career and on his ability to maintain a broad range of intellectual and human interests throughout his long life, is in itself a bright example that this ambitious aim can be pursued, and even accomplished.

I wish to dedicate this article to the memory of my mother, Letizia Morabito Rubino (1929-2009), a passionate librarian and teacher who strongly encouraged me to have my Italian version of Cantor's Dilemma *published and to participate in the 2009 Dortmund symposium on Carl Djerassi shortly before her passing away.*

Works Cited

Aldston Jr., Alfred N. and Dina L.G. Borzekowski, Jonathan A. Eisen, Sheri L. Fink, E. Weber Hoen, Dean Y. Hung, Shirley Lin, Cynthia T.M.H. Nguyen, Julie E. Phillips, Michelle Stohlmeyer, Cenk Sumen, Craig A. Swanson, Noriko Takiguchi, Yvonne Thorstenson, Harriet A. Washington. 1998. "A science renga." *Nature* (London) 393, 11 June: 512-13.

Angier, Carole. 2002. *The Double Bond: Primo Levi – A Biography.* London: Penguin.

Cham, Jorge. (2009). http://www.phdcomics.com/comics.php

Coelho, Sara. (2009). "Piled Higher and Deeper: The Everyday Life of a Grad Student." *Science* 323: 1668-69.

Djerassi, Carl. 1989. *Cantor's Dilemma. A Novel.* New York: Penguin.

Djerassi, Carl. 1998. "Ethical discourse through science-in-fiction." *Nature* (London) 393, 11 June: 511.

Djerassi, Carl & Roald Hoffmann. 2001. *Oxygen.* Weinheim: Wiley-VCH.

Djerassi, Carl. 2001. *This Man's Pill: Reflections on the 50th birthday of the Pill,* Oxford: Oxford UP.

Djerassi, Carl. 2005. "Meeting strangers and three autobiographies." www.webofstories.com/play/16823.

Djerassi, Carl. 1991. *The Clock Running Backwards.* Brownsville, OR: Story Line Press.

Djerassi, Carl. 1994. *Bourbaki's Gambit.* New York: Penguin.

Djerassi, Carl. 1990. *Steroids made it possible.* Washington, DC: American Chemical Society.

Editorial. 2009. "Summer reading: science in fiction." *Nature Methods* 6/7: 471.
Hensarling, Jeb. 2011. http://thinkprogress.org/special/2011/07/29/283614/jeb-hensarling-debt-dna/?mobile=nc
Hoffmann, Roald. "Men and Molecules." www.roaldhoffmann.com/pn/modules/Downloads/docs/Men_and_Molecules.pdf
Lablit. 2009. www.lablit.com
Levi, Primo. 1985. "Ex-chimico." In: P.L. *L'altrui mestiere*, Torino: Einaudi. 12-14.
Pallavicini, Piersandro. 2005. *Atomico Dandy*. Milano: Feltrinelli.
Prasher. 2008. www.nytimes.com/2008/10/17/science/16prasher.html
Prasher. 2011. http://discovermagazine.com/2011/apr/30-how-bad-luck-networking-cost-prasher-nobel/article_print
Rohn, Jennifer. 2006. "Experimental fiction." *Nature* (London) 439, 19 January: 269.
Rohn, Jennifer. 2010. "More lab in the library." *Nature* (London) 465, 3 June: 552.
Sokal affair. 2011. http://en.wikipedia.org/wiki/Sokal_affair
Tomatis, Renzo. 1963. *Il laboratorio*. Torino: Einaudi.
Tomatis, Renzo. 2011. http://omega.twoday.net/stories/235689
Williams, Dudley H. & Fleming Ian. 1995. *Spectroscopic Methods in Organic Chemistry*. 5th ed. London, New York: McGraw-Hill.

Science-in-Poetry?
Chemistry and the Metaphysical Tradition

Walter Grünzweig

In 2012, Carl Djerassi published a bilingual book of poetry that wisely omits the genre in its title. Hoping for greater success and attention than poems might command, the volume defines itself as a diary, namely *The Diary of Pique. 1983-84*. The occasion for the book, the author tells us in one of the several extended prose sections, was a – temporarily – unhappy relationship in the years of 1983-84 (according to Barbara Ehrenreich part of the "worst years of our lives"): his lover Diane Wood Middlebrook, a prominent Stanford English professor and biographer (among others Ted Hughes, Sylvia Plath, Ovid), had left him.

Throughout the history of humankind and poetry, lovers have been inclined towards poetry. Those who have newly fallen in love write poems, often bad ones because of an all-too-immediate translation of fiery emotions into text. Sometimes they employ highly conventional metaphors, mostly derived from nature imagery, to express the urgency of their feelings. *Un*happy lovers, too, like to write poems, celebrating their pain and their sorrow in verse and rhyme.

Not so the speaker in these poems. His poems do not express sorrow; they convey his rage. Pain can only be indirectly gathered from these lines. The speaker is mad. His lover has left him, and he is full of *pique:*

The pain is mostly gone now,
Displaced by resentment.
Rich, fertile, ever-swelling resentment.
Not a pretty feeling,
But neither is revenge,
Especially when born of resentment. (34)

The frankness of these negative emotions is surprising, as is their threatening nature:

I shall write you poems,
Public poems for all to read,

Where the private meaning is the lock
To which only you carry the key.
If keys can be bitter,
Let this be one. (32)

This is not meant in the sense that the general public would read these texts as poems only and the specific addressee, the departed lover, would get their 'true' meaning. These texts clearly spell out the identity of the speaker even without the explanatory prose or the author's name on the title page. Carl Djerassi frequently complains in interviews that journalists are more interested in his reputation as the father (or "mother," as he prefers) of "the Pill" than in his literary *oeuvre*, and he actually does much to upgrade the significance of his writing vis-à-vis his fame in chemistry. The speaker of these poems, however, very much highlights not only his identity as a chemist, but also specifically as the creator of the Pill. In the wonderful poem entitled "The Clock Runs Backward," where "the man who has everything" receives a special clock taking him back through time, he clearly marks this achievement:

What about twenty-eight?
Ah yes – he nearly forgot.
The year of THE PILL.
The pill that changed the world.
No – too pretentious, too self-important.
But he did change the life of millions,
Millions of women taking his pill, he thought. (18)

In another poem, he calls himself a "master of chemical mutations, / Whose alchemy touched millions." (90) Less explicitly, but with a pose that is no less self-assured:

The middle aged, yet age-oblivious Man,
Ostensibly domineering,
Attractively persuasive,
Intriguing in his affluence and taste,
With flashes of personal vulnerbuility,
In one of his several roles
Acts to the fullest the professor
And meets a student. (134)

The German translation, printed on the right-hand side in this attractive bilingual edition and more than authorized by the poet, specifies that this student is a "Studentin," a female. The speaker of these poems thus addresses the biographical reality of the author, and the association of qualities of the speaker – such as business success, scientific ingenuity, affluence, womanizing – with that of the author is obviously written into the text.

This is not done without self-criticism and self-irony. These poems, in contradistinction to Djerassi's autobiographies, his fiction, and his plays, show what is *beneath* this person – beneath his clothes but also beneath his skin. In the cycle of poems entitled "Hairshirt," the speaker treats himself with ruthless candor:

Arrogant – often barely tolerable;
at times becomingly; ...
Age – past the prime;
(he thought not too noticeably),
Hair – silver waves;
(his narcissistic synonym for grey). ...
Anything else of relevance?
Not really, except his loneliness.
So overpowering,
So persistent,
As to be overlooked by most
(Including her). (39f.)

The poems even address the greatest tragedy in Djerassi's life, the suicide of his daughter, connecting it, in the following lines, with the departure of his lover:

Do I want to celebrate my birthday
With champagne, four hundred guests,
Ambassadors?
No, as usual, I am more demanding.
Not with four hundred, but with two.
Two women,
One dead, the other gone,
Each as she chose.

I shall celebrate alone. (84-86)

This very touching moment dramatizes to the extreme the loneliness frequently expressed in these poems. The lover, too, is referred to with explicitness and extreme candor. She is

A littérateur, poet and wordsmith,
with barely an use for scientists
Highly honed prose – almost too polished.
Flawless reader – a true Vocalissima.
Self-assured – to all except her shrink. (40)

Diane Wood Middlebrook can easily be identified professionally; at the same time, her deficits are pointed out – as if the speaker wanted to find reasons to be able to disassociate himself from her. In rendering her body, an ambivalence appears that is characteristic of the whole text:

Not truly beautiful:
hips: a touch too high;
breasts: full, but now a touch too droopy;
forehead: a touch too short;
nose: a touch too small;
hair: a touch of grey needing to be masked;
legs: spectacular, with just a touch of blemish. (40)

Ostensibly, this lyrical strategy might help the speaker to distance himself from her and to cope more successfully with his loss. After all, she is not *that* beautiful, there are more beautiful, more perfect women in this world. But the strategy fails and indeed results in the opposite because these "imperfections" are "[B]lended into a startling presence." (42) His main problem remains: "[S]he was only the most arresting woman he'd ever met ... In other words / **Avis rarissima**." (42)

It is not quite as easy to identify the name of his rival from these poems, but the repugnance of the speaker is real and so is his venom:

May your new lover break out in hives
At your gentlest, scented caress.
May drenching, juicy beads
Of reeking, oily sweat
Spread all over his horny skin.
Then think of me. (106)

Read from this angle, this poetry resembles the celebrity reporting in the so-called "news" pages on the internet, in tabloids and, increasingly, in quality journalism. In fact, the exhibitionism and voyeurism in Djerassi's poetry seems more profound because it not only reveals relational and sexual constellations but lays bare the bottom of the speaker's psyche: the speaker strips down to his heart.

It would, however, be unfortunate to limit oneself to the biographical reading of these poems – although the reader is virtually seduced to engage in this mode of reception. I would like to suggest that this seduction is part of the textual strategy and that readers have to resist it in order to appreciate not just the poems as poems but also as meta-lyrical commentaries on poetry. The question that needs to be answered is whether these poems can stand for themselves and by themselves – in spite of the prominence of their author in the 'world.'

Has Carl Djerassi, following his invention of two generic terms, science-in-fiction and science-in-theatre, now also given us "science-in-poetry"? Djerassi has not yet used that term, but I would like to claim that the latter is more to the point than the earlier two. Whereas in fiction and theatre, science is essentially a *theme* to be investigated, the two areas actually intersect, interact, and are fused in poetry.

The figure of speech I would like to employ here to explain Djerassi's poetry is the *conceit*, originally derived from poetry in the Romance languages but accessible to Djerassi mostly through the English metaphysical poets of the 16th and 17th centuries (John Donne's well known line "No man is an island, entire of itself" actually introduces Djerassi's poem "My island," 76). A conceit (as even undergraduate students of English are well aware of) "establishes a striking parallel – usually an elaborate parallel – between two very dissimilar things or situations." (Abrams 32) In Helen Gardner's more disrespectful words, a conceit is "a comparison whose ingenuity is more striking than its justness or at least is more immediately striking. All comparisons discover likeness in things unlike." (Gardner 19)

The metaphysical poets exploited all knowledge – commonplace or esoteric, practical, theological, or philosophical, true or fabulous – for the vehicles of these figures; and their comparisons, whether succinct or expanded, were novel, witty, and at their best startingly effective. (Abrams, 33)

The conceit joins two areas of reality in one poetic image. What is most significant about this pre-romantic image in our context is that its 'understanding' is not so much a function of human intuition, feelings and emotions but of our reason. Metaphysical poetry develops a poetic *argument*. It makes demands on the reader's intellectual capacity.

In much of Djerassi's *Diary of Pique*, two "dissimilar" areas, chemistry and human relationships (friendship, partnership, love, sexuality), are joined together. At first, this does not seem so new. There is a very important term in everyday language that connects these two areas, namely "relationship chemistry." The internet, the best medium to study the state of human intimacy, gives great prominence to the concept. Among the hundreds of explanations found, relationship chemistry may mean that "You feel a spontaneous connection with each other from the very beginning" ("Is Chemistry in Relationships Important"); that there is a spiritual affinity that confirms and maintains the physical relationship, "physical chemistry" being important but not everything (see Coleman); or that it is complemented by "romantic," "social," and "sexual chemistry." ("Relationship Chemistry")

In this concept, the *existence* of a "chemistry," and not the successful *combination* of two substances, is seen as a prerequisite for the beginning of and a sine-qua-non for a continuing sexual relationship. It is a somewhat sloppy, albeit extremely successful metaphorical use of the science of chemistry, probably mostly related to the inexplicability of the success of a relationship (which is more characteristic of alchemy, the precursor of chemistry) or to its spontaneity and its serendipitous, experimental quality (which is indeed a quality of chemistry as we know it today).

Djerassi's use of his discipline as metaphor is much more calculated and rational – precisely in the tradition of the metaphysical poetry I have referred to above. In a poem entitled "Push-Pull," he uses a chemical concept in the hope of a reversal of his fortune:

I, the chemist, know all about
The push-pull mechanisms
Of chemical reactions.

You, the long-time collector of men,
Should know all about
The push-pull of love:

The pull of the new lover,
The push from the old.
But beware!

Push and pull differ but in one direction.
All that's required is
To turn. (92)

This truly exceptional poem is strongly reminiscent of one of the best known metaphysical poems, John Donne's "A Valediction: forbidding mourning," which uses the image of a drawing compass to express the endurance and intensity of a separated couple's love. The power of the poem depends on the ability to understand the workings of a compass and to transfer the paradox of distance and proximity to the position of the lovers.

If they be two, they are two so
As stiff twin compasses are two,
Thy soule the fixt foot, makes no show
To move, but doth, if the' other doe.

And though it in the center sit,
Yet, when the other far doth rome,
It leanes, and hearkens after it,
And grows erect, as it comes home.

Such wilt thou be to mee, who must,
Like th'other foot, obliquely runne;
Thy firmness makes my circle just,
And makes me end where I begunne. (Gardner 74)

The essential argument is made in the final lines. In Djerassi's poem, too, the end of the poem, a tercet, must provide an explanation that both consoles the speaker and convinces him of the possibility of a restored relationship. "Push and pull," in itself a formula with a strong sexual connotation, provides the basic structure of the erotic relationship – which, however, is not monodirectional. Without stopping the movement and without a fundamental change, a turn will bring about a very different, indeed opposite, constellation – and return the lover back to the original partner.

The insight the chemical model in Djerassi's poem provides is *new* and surprising, and, more importantly, plausible. But it not only implies that a law in the natural sciences might also be valid in the system of human relationships. The idea, that love is a force that can easily change direction, is something that is believable in itself – and becomes graphically comprehensible through this example from chemistry.

The difference between the two authors is that whereas Donne, a theologian and poet, used geometry metaphorically in this, his most famous poem, in Djerassi, the metaphor comes from his own professional life. Whereas chemistry remains the vehicle, it has a more autonomous status. This becomes even more obvious in a poem entitled "You wash this shirt like a chemist":

I didn't know there were *chemist's hands,*
Do I touch you like a chemist?

Grip your wrist
The way I grip the necks of Erlenmeyer flasks?

Hold your buttocks
The way I hold round-bottomed vessels?

What else does this chemist do?

Suck your tongue
The way I suck pipettes?

Carry your night scent in the morning
The way I carry my day's lab odor in the evening?

Wear your bath robe
The way I wear my lab coat?

Is a chemist's alchemy permitted in the bedroom?
The literature is strangely silent on this topic.

Yet if anyone knows,
You do. (124)

Although the title nominally makes a comparison – you behave *like* a chemist – the speaker in this poem *is* a chemist, and the two parts of the metaphor relating the chemist and the lover are therefore situated on the same plane.

The poem does not display a central conceit like the previous one. Yet, it does require careful consideration of the work and life(style) of a chemist in order to understand the tenor, and trying to spell it out will likely turn the interpretation into a manual for successful sexual practices. The point of origin of the situation, placed in quotation marks in the poem's title, was probably the lover's comment watching the speaker wash his shirt. What could she have meant by this comment? That he is overly careful, that he touches the material in a certain way, that he is too conscious of the detergent he is using? That he takes too long, that he is too careful, too involved in the process of 'merely' washing a piece of clothing?

What exactly she meant is not explained, and the speaker probably never finds out. Rather, the question is the basis for a series of speculations about the meaning of the speaker's chemical profession for his private life, for his relationship with the questioner, especially in sexual terms. The first couplet begins *in medias res* but does not yet refer to any concrete physical action. "Do I touch you like a chemist?" of course raises the question whether the lover can discern any difference between her previous, probably non-chemical lovers and himself. Is the chemist more attentive to details? Does his professional grasp of material substances make him a specialist in dealing with a female body?

The second couplet seems at first almost comical. Erlenmeyer flasks remind many of us of our chemistry classes in school. The flask's shape – the wide, conical bottom and the narrow, cylindrical neck – are reminiscent of a human body, especially that of a person well endowed around the hips. The advantage of the Erlenmeyer flask is, of course, that because of its shape there is less of a risk of spilling. One can handle it with greater energy and shake it without fear of losing the contents. Gripping his lover's wrists like an Erlenmeyer flask means to approach the sexual partner with a certain degree of intensity, excitement, even a bit of aggressiveness. In the next couplet, its shape is directly and explicitly addressed: holding the buttocks like one would hold a round vessel, in firm grip, with both hands, and understanding the significance of the fluid(s) contained therein. Similarly, the idea of comparing the sucking of a tongue to that of pipettes has a strongly physical, technical but also an expert quality, which gives the erotic action a fascinating concreteness and directness.

The three subsequent couplets operate more traditionally on the metaphorical level, exploring the relationship between professional and private lives. Yet, in juxtaposing the bath robe (of his lover) and the lab coat, interesting connections come up from the potentially erotic attitude the chemist has towards his work to the professional enthusiasm he may transfer to his private (sexual) life. As we have seen above, the speaker, asking the question "Is a chemist's alchemy permitted in the bedroom?" is wrong when he answers it by saying, "The literature is strangely silent on this topic." (124) The internet is full of "relationship chemistry" in the bedroom, but it is not the "alchemy" the speaker is referring to. Rather, we may assume he is talking about the medieval alchemist's quest, which is transferred into the bedroom. Is there a place for this mixture of the scientific and the esoteric in the bedroom? The answer must be provided by the (absent) lover: "Yet if anyone knows, / You do." (124)

There are other, more indirect hints at the "chemical" approach to life, which are nevertheless highly relevant to the poetry, as in a poem entitled "Bodybuilder":

You, the body-builder,
The one with the glutes,
Surrounding a deep, caressable valley;

You, the one with the pecs,
Supporting small, perfect diamond nipples;

You, the one with the lats,
Bi's, tri's, abs and quads,
All of them glistening;

You, whose eyes sparkle,
Whose voice laughs,
You cried last night.

Do you know
That a body-builder's tears
Flow differently,
Taste differently
From those of other women? (120)

There is an intertextual connection here to one of Djerassi's earliest texts, a short story entitled "The Dacriologist," where a chemist claims he can tell the difference between tears produced by sorrow and tears coming from laughter. When he is put to the test, he is puzzled because a woman's samples fit neither category. The explanation is as simple as it is surprising: the tears examined are the result of a female orgasm.

As in the short story, the question is not how they flow differently or how they taste differently, though that, too, might be of interest. The point to the poem is the surprising fact that somebody licking the tears of his lover (tears that might very well have had coital origin) might actually try to classify their tastes and differentiate them. The chemist's approach to his lover, especially when the latter is a body-builder, is rather novel indeed.

As stated initially, Djerassi claimed that his frustrated love was the origin of his poetry:

My desire for revenge turned into an outpouring of poems – confessional, self-pitying, even narcissistic. It was a cathartic experience for someone who until then had never written a single line of verse – cathartic because I wanted to revenge myself on her own turf and that of her new lover, who was not a scientist but a literatus manqué. With a few exceptions, none of these poems was published or read by anyone else. They simply had turned into the diary of an unhappy, revengeful man who never before (or since) had kept a diary. (14-16)

This is not entirely true. In 1991, Djerassi had indeed already published a small volume of poetry entitled *The Clock Runs Backward* (without any prose commentaries) of which the large majority found their way into *The Diary of Pique*. Moreover, as indicated in the earlier volume, many of those poems had previously been published in journals. The point to be made here is that Djerassi has partially modified his poetry to fit his revenge story. The "My muse is resentment, the poem my revenge," from which I earlier quoted the lines "The pain is mostly gone now, / displaced by resentment," concludes with the following passage:

How does a chemist
Revenge himself against a poet?
Synthesize a poem?

Distil its essence?
Filter the impurities?
Evaporate it to dryness?

Stop the sophistry!
Write the poem. (34-36)

In *The Clock Runs Backward*, these lines alone form the introductory poem, entitled "Exhortation," to the volume. The first two lines, however read:

How does a chemist
Transmute himself into a poet?

I am pointing out this significant variant to show that Djerassi's synthesis of chemistry and poetry also leads to a remarkable degree of poetic self-reflexivity. In such poems, the scientist reflects the process of invention and innovation. What can we learn from chemistry to understand the process of creativity? Are the reactions of compounds leading to synthesis, i.e. the bringing together of different substances to produce something new comparable to the assemblage of different experiences, ideas, and feelings to produce poetry? What, then, is the equivalent of the chemical reaction?

Is, to move to the next line, the slow process of poetic maturation (at least in some poets, including one of Djerassi's favorites, Wallace Stevens, to whom a whole chapter is dedicated in *The Diary of Pique*) equal to a process of distillation to get to a poetic "essence"? Similarly, the processes of filtering and evaporation share the notion of developing an improved text whose high or special quality is developed through processes of rarefication. It is a notion of literature insisting on extreme linguistic economy. According to Helen Gardner, "concentration," the fact that the "reader is held to an idea or a line of argument," is central to metaphysical poetry. (17)

The most important meta-lyrical poem in this collection is entitled "Why are chemists not poets?" At first, the parallels between chemical and poetic models are shown:

Take the chemists.
By definition
Synthesizers of molecules;
Dissectors of molecules;
Manipulators of molecules.

For *molecules* read *words*:
You have defined the poet. (60)

It is interesting that molecules to Djerassi are metaphorically operative on the lexemic and not, for example, on the phonemic level. Especially in 20th century experimental poetry, the phoneme is at times as significant to 'poetic synthesis' as the word. He then goes on to problematize the analogy:

Why then
Are chemists seldom poets?

Initially
Chemists dilute
Yet eventually
They always concentrate,
Evaporate, still
To reach their goal. (60)

The previous parallel between chemist and poet is now questioned because

While some poets dilute,
The best a thickeners,
(No wonder the German poet is a *Dichter*.)
They thicken, *dichten*, concentrate,
Distill
Until the poem is compacted. (60)

This etymology is mixed up. The German word *dicht*, etymologically related to English *tight*, has a different origin than *dichten*, which came into the Germanic languages from Latin *dictare*. But there seems no fundamental difference between the processes described in the first part for the chemist and in the second section for the poet. Rather, the skillful handling of substances and their manipulation – processes in time – are the essence of both processes.

From this uneven comparison, the poem moves on to the character of the actors, chemists and poets, themselves:

Why then
Are chemists rarely poets?
Chemists peer through safety glasses,
Working in fume hoods,
Behind explosion-proof shields,
In a partial vacuum or under an inert atmosphere.

Can you imagine a poet
Writing in such a verse laboratory?
Writing sterile,
Non-explosive,
Non-inflammable,
Vacuous poems?
Poems which, if they stink,
Are kept in a rarefied atmosphere
To hide the stench?

Now you know why
There is no future
For a careless chemist
A cautious poet. (60-62)

The difference between the two professions – or vocations – are the conditions under which the work is carried out. Whereas the chemists are subjected to a strict régime of safety, poets have more freedom, more room for their work. But these differences, which may be significant in that they describe the opposition of materiality and intellectuality, cannot hide the fact that the comparison highlights that both are engaged in experiments designed to produce innovation and novelty. Indeed, he does not say that poems are not explosive or inflammable, but that the poet does not have to take the measures of precaution that are standard for the chemist.

With his lyrical juxtaposition of chemistry and poetry, Djerassi has made a truly original and creative contribution to the field and science and literature at large. Not limited to the thematic and biographical levels, structural similarities, even homologies, appear which are instructive and can help us on our way to recreate the common foundation of the arts and the sciences (which was still taken for granted by the metaphysical poets).

Of course, not all of Djerassi's poetry, either in *The Clock Runs Backward* or in *The Diary of Pique* have chemistry as the master metaphor. Much would have to be said about Djerassi's relationship to Wallace Stevens and to Paul Klee and the inter-arts dimensions and contexts of his poetry. But most of Djerassi's poems share, with the chemical poems discussed, the anti-Romantic quality of intellectual poetry which nevertheless (as metaphysical poetry also does) has an immediate and strong effect on the senses and on human sensibility:

The faces of models in fashion magazines
No more reflect the moods of the models
Than the skin of the mango
Gives you a test of its flesh. (74)

Works Cited

Abrams, M. H. 1993. *A Glossary of Literary Terms.* 6th edition. Fort Worth et al.: Harcourt Brace Jovanovich.
Coleman, Toni. 2012. "True Love and Chemistry: Exploring Myth and Reality." www.enotalone.com/relationships/2946.html
Djerassi, Carl. 1991. *The Clock Runs Backward.* Brownsville, OR: Story Line Press.
Djerassi, Carl. 1988. "The Dacriologist." In: C.D. *The Futurist and Other Stories.* London: Macdonald. 29-46.
Djerassi, Carl. 2012. *Tagebuch des Grolls. A Diary of Pique. 1983-1984.* Innsbruck & Wien: Haymon.
Gardner, Helen. (1966). *The Metaphysical Poets.* Rev. ed. Harmondsworth et al.: Penguin.
"Is Chemistry in Relationships Important?" 2012. www.mens-relationship-advice.com/chemistry-in-relationships.html#axzz1nWJHOPqT
"Relationship Chemistry." 2012. In: *Encyclopedia of Human Thermodynamics, Human Chemistry and Human Physics.* www.eoht.info/page/Relationship+chemistry

Arts and Translation

Multiplizität als künstlerische Strategie: Zum wissenschaftlich-künstlerischen Fotodialog von Gabriele Seethaler und Carl Djerassi

Carl Aigner

Ich ist ein anderer.
Arthur Rimbaud

Kein anderes Bildmedium hat das Thema Portrait derart proliferiert wie die Photographie. Das gesellschaftliche Bedürfnis nach einem Selbstbildnis als existentielle Selbstvergewisserung ist seit dem 19. Jahrhundert eine Ultima Ratio dieser bildnerischen Sehnsucht geworden. Das photographische Bild erfüllt dieses neue Bedürfnis für die bürgerlich-industrielle Gesellschaft in kongenialer Weise: Sie ist damals die schnellste Möglichkeit, von sich Bilder gewinnen zu können, von den Kosten her wurde es im Laufe des 19. Jahrhunderts zunehmend bewältigbar, sie hat einen neuen Zeit- und Wirklichkeitsbezug und bietet auch eine neue Form der Vervielfältigung.

Es ist die neue bildtechnische Multiplizität von Negativ und Positiv, die das bildnerische Prinzip der Serie konstituierte, welches etwa Andy Warhol so famos zur Grundlage vieler seiner Arbeiten machte. Nicht der Kubismus, wie immer gesagt wird, sondern die Erfindung der Photographie ist es, welche das zentralperspektivische Bild der Renaissance dadurch radikal aufzubrechen begonnen hat. Wiewohl bereits Ende des 17. Jahrhunderts zentralperspektivisch das Mehrfachbild mit Front- und Seitenansicht in einem Bild entwickelt worden ist, inauguriert die Photographie den Beginn der Polyperspektivität, indem sie die Serialität des Bildes und durch die Möglichkeit der Mehrfachbelichtungen die Bildtechnik der „Collage" begründet: Die Multiplizität ist neben dem „zertifikatorischen" Moment (Roland Barthes) das eigentliche Wesen des photographischen Bildes.

In subtiler Weise spürt die Biochemikerin und Photographin Gabriele Seethaler dem multiplen Denken der Naturwissenschaftler nach, indem sie diese neuen pikturalen Möglichkeiten des photographischen Bildes auslotet und mit ihnen verknüpft Von Beginn an stand dabei die Frage nach der pikturalen Identität im Mittelpunkt ihres Interesses. „Identität genotyp-phaenotyp" betitelt sie eine große Bildserie, bei der das Portraitbild mit einem genetischen Fingerabdruck (Prof. Franz Neuhuber) und der musikalischen Identität (Komponist Renald Deppe) verschmolzen wird. Rasch wurde das einförmige Portrait in neue wissenschaftliche Erkenntnisse transferiert und das unilineare Photobild aufgebrochen.

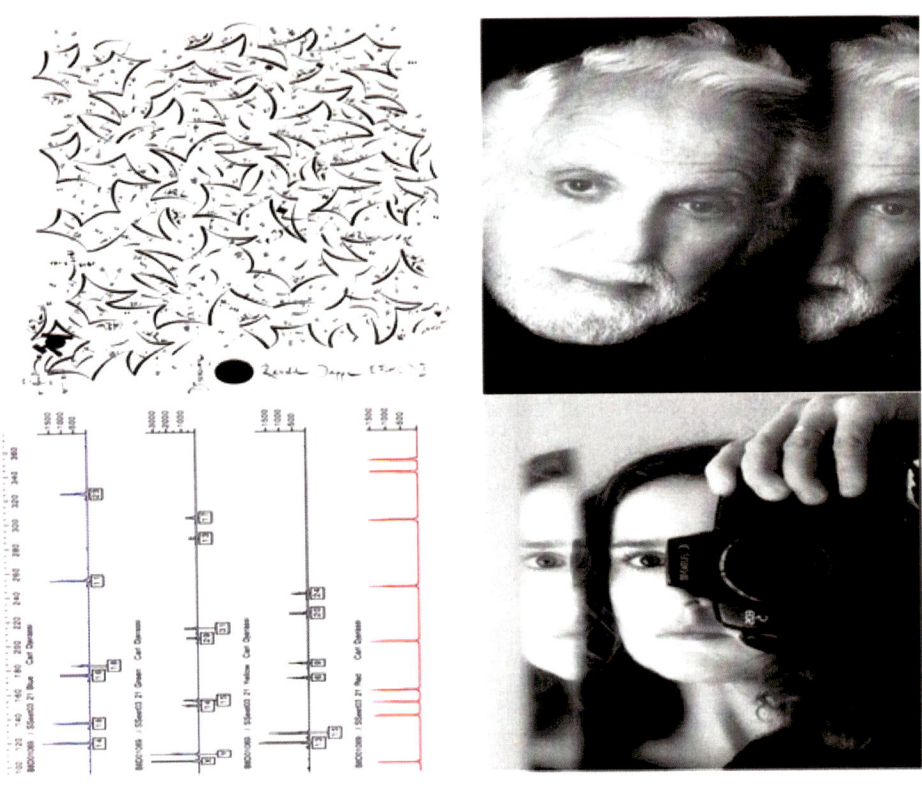

Abb. 1: Gabriele Seethaler in Zusammenarbeit mit Renald Deppe und Franz Neuhuber: Identität genotyp-phaenotyp, Carl Djerassi, 2004; Selbstportrait Gabriele Seethaler, 2000

Neben der Mehrfachbelichtung als simultane Parallelität werden facettengeschliffene Spiegelbilder zu einem Ort des Mehrfachbildes als ein Bild. Mitten in dieser für sie neuen engagierten Tätigkeit als Photographin lernt sie 2004 Carl

Djerassi kennen und beginnt 2006 auch mit ihm eine über viele Monate währende portraitistische Auseinandersetzung. Doch rasch merkte sie, dass diese Form der Bildduplizität nicht ausreicht, um die philosophischen, literarischen und naturwissenschaftlichen Dimensionen des Denkens von Djerassi photobildnerisch adäquat zu erfassen.

Die Leidenschaft Djerassis für das Werk von Paul Klee führt ihn zu Walter Benjamins Reflexionen über das Aquarelle „Angelus Novus" und zur Auffassung von Konrad Eberlein (Universität Graz) und Geoffrey H. Hartman (Yale Universitiy), die unabhängig voneinander erstmals darauf hingewiesen haben, dass Paul Klee Adolf Hitler als den „Angelus Novus" bildnerisch wahrscheinlich vor Augen hatte, als er dieses Werk bereits 1920 am Bauhaus in Weimar realisiert. Diesen, von Djerassi in seinem Buch und Dokudrama *Vier Juden auf dem Parnass. Ein Gespräch* aufgenommenen Standpunkt versucht Seethaler durch Überarbeitung der Hitlerphotographie mit erhobenen Armen darzustellen. Mittels der Lentikulartechnik gelingt es ihr, in einem Bild – je nach Blickwinkel des Betrachters – den Angelus Novus oder Hitler mit den erhobenen Armen, beziehungsweise Übergangsformen der beiden in einer pikturalen Verschmelzung sichtbar zumachen.

Benjamin, der das Werk besaß, sah in ihm den „Engel der Geschichte" und schrieb in den Thesen „Über den Begriff der Geschichte" im Jahr 1940: „Er hat das Antlitz der Vergangenheit zugewendet. Wo eine Kette von Begebenheiten vor uns erscheint, da sieht er eine einzige Katastrophe, die unablässig Trümmer auf Trümmer häuft und sie ihm vor die Füße schleudert. Er möchte wohl verweilen, die Toten wecken und das Zerschlagene zusammenfügen. Aber ein Sturm weht vom Paradies her, der sich in seinen Flügeln verfangen hat und so stark ist, dass der Engel sie nicht mehr schließen kann. Dieser Sturm treibt ihn unaufhaltsam in die Zukunft, der er den Rücken kehrt, während der Trümmerhaufen vor ihm zum Himmel wächst. Das, was wir den Fortschritt nennen, ist dieser Sturm." Gleichzeitigkeit des Ungleichzeitigen und diametrale Polyperspektivität kennzeichnen dabei die interpretatorische Sicht Benjamins auf das Aquarell von Klee.

Es ist die These beider Wissenschaftlicher, die Djerassi zur visionären geschichtlichen Sicht von Klees Angelus Novus brachte und die Seethaler animierte, diese in „neuer" Weise in einem Bild zu amalgamieren. 2008 greift sie erstmals dabei auf die fast vergessene lentikulare Bildtechnik als Verfahrens- und Darstellungsweise zurück, nicht zuletzt auch auf Grund von Kindheitserinnerung an diese besondere Bildmöglichkeit. Diese in den 1940er Jahren in Amerika popularisierte Bildtechnik wurde in den 1960er und 1970er Jahren auch in Österreich zu einer beliebten Bildmöglichkeit. In experimenteller Weise wird damit ein photographischer Approach realisiert, der das Denken von Carl Djerassi gewissermaßen photobildnerisch transzendiert.

Abb. 2: Gabriele Seethaler, 2006, Angelus novus benjaminianus var.3,
Originalfoto Walter Benjamin auf Mallorca, Fotograf unbekannt;
Walter Benjamin Archiv, Akademie der Künste, Berlin.

Die Frage nach der (sexuellen) Identität im Zeitalter der Gen- und Biotechnologie in den literarischen Überlegungen von Djerassi und die Erfahrungen mit dem

Multiplizität als künstlerische Strategie

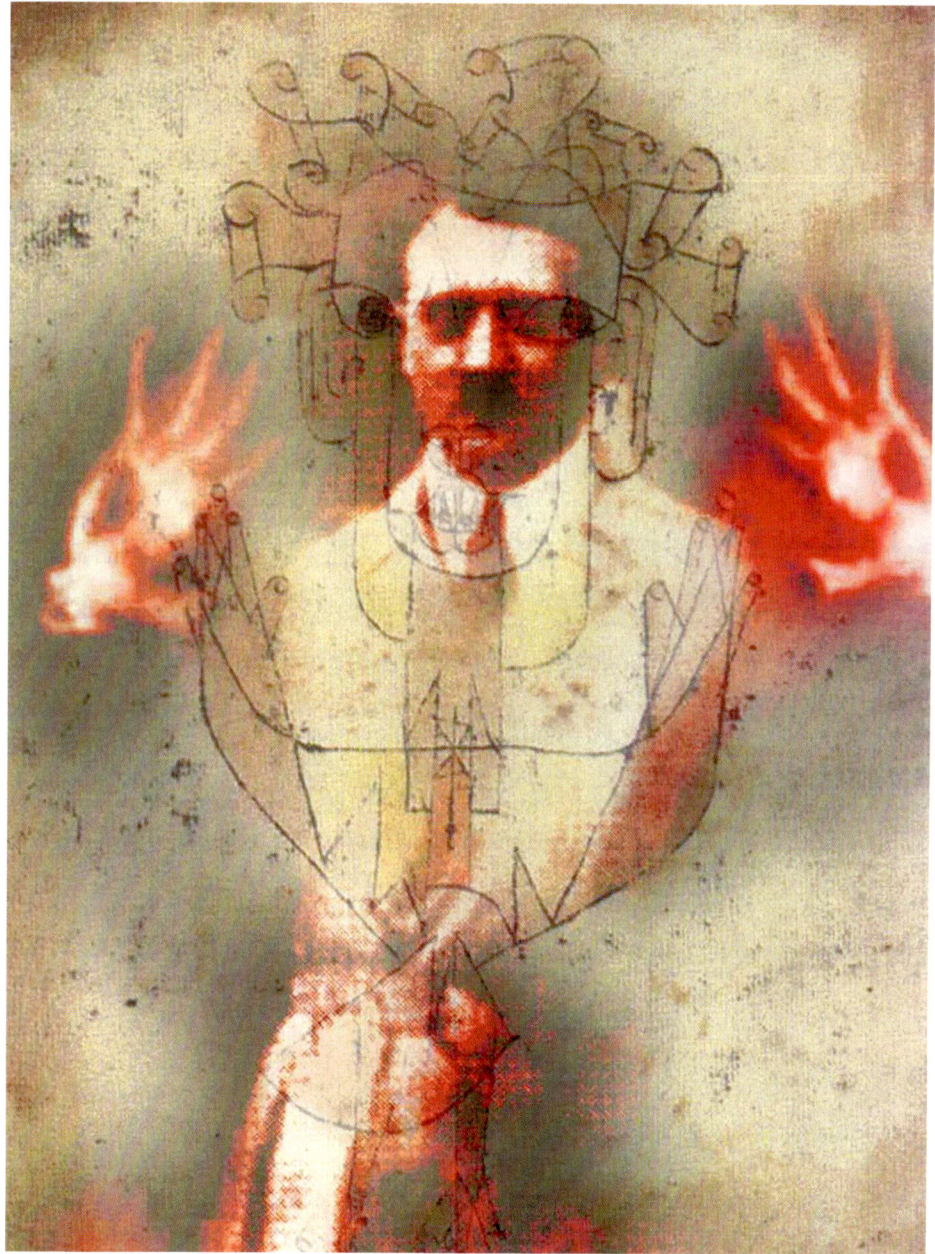

Abb. 3: Gabriele Seethaler, 2006, Transformation Paul Klees Angelus novus – Hitler.

Werk von Klee führten für die Photographin zu einer vielfältigen und komplexen Auseinandersetzung mit grundsätzlichen Fragen des Portraits und der bildnerischen Darstellbarkeit des Menschen, in der die Lentikulartechnik eine entschei-

Abb. 4: Gabriele Seethaler, 2006, Collage Angelus Novus – Engel der Geschichte; nach Paul Klees Angelus Novus und Walter Benjamins Thesen „Über den Begriff der Geschichte",1940.

dende Rolle einnimmt. Sowohl in ihren „Generationsportraits", in den Identitätsbildern ihres „Genotyp-Paenotyp-Projekts" als auch in ihren dreidimensionalen zellulären Bildern ist die Möglichkeit einer Polyperspektivität grundlegend geworden.

„Vor dem Objektiv bin ich zugleich der, für den ich mich halte, der, für den ich gehalten werden möchte, der, für den der Photograph mich hält, und der, dessen er sich bedient, um sein Können vorzuzeigen", beschreibt der französische Philosoph Roland Barthes die Komplexität von photographischen Portraitbildern. Seethaler erweitert mit ihren Arbeiten den Blick in die Vielschichtigkeit der personalen Identität im 21. Jahrhundert und stellt damit erneut die Frage nach dem Subjektbegriff und der humanen Identität im Zeitalter genetischer Reproduzierbarkeit, wie sie so eindringlich Djerassi in seinen Romanen und Theaterstücken vermittelt. Seethaler demonstriert aber auch nachdrücklich, dass der Subjektbegriff einer Epoche und Gesellschaft immer unmittelbar mit deren Bildbegriff korreliert, wenn Bilder eine so fundamentale Bedeutung bekommen, wie dies in der okzidentalen Gesellschaft und Kultur seit der Renaissance der Fall ist.

Noch in den 1970er Jahren galt der Begriff der multiplen Persönlichkeit als Beschreibung eines psychischen Krankheitsbildes. Heute ist es angesichts einer derart komplexen und medial konditionierten Gesellschaft ein Positivum geworden. Mehrere Jobs auf einmal, mehrere Identitäten gleichzeitig etwa durch das Internet, führen zu einer polyperspektivischen Persönlichkeitsformulierung, die schon mit den Erkenntnissen eines Sigmund Freud evident geworden sind und in der Literatur etwa in Form des Doppelgängermotivs seit dem 19. Jahrhundert en vogue ist. Seethaler hat in ihrer photographischen Auseinandersetzung mit Werk und Person von Carl Djerassi eine bildnerische Perspektivität (weiter-)entwickelt, in der und durch die die Diversität des Subjekts visuell erfahrbar wird: Wenn das Ich ein Anderer ist, wie Rimbaud zu Beginn des 20. Jahrhunderts schrieb, dann sind wir Mutanten dieser Diversität: Bild-Schamanen des 21. Jahrhunderts.

Zitierte Literatur

Benjamin, Walter. 2007. „Über den Begriff der Geschichte." In: W. B., *Schriften zur Theorie der Narration und zur literarischen Prosa.* Frankfurt/M.: Suhrkamp. 129-140.

Abb. 5: Carl Djerassi und Gabriele Seethaler 2006, Collagen nach Paul Klees Geschrei, Noch weiblich, Gruppe zu elf, Brandmaske, Engel übervoll, Sirene mit der Altstimme.

Abb. 6: Carl Djerassi und Gabriele Seethaler, 2006, Paul Zion Klee (Paul Klees jüdische Symbole überlagert von Paul Klees Gesicht). Originalfoto Paul Klee, Dessau, 1927: copyright Hugo Erfurth, Standort: Archiv Bürgi, Bern.

Carl Djerassi und Paul Klee:
Ein Dramolett

Klaus Albrecht Schröder

D: Sprechen wir also über Klee.
K: Nicht schon wieder. Er beschäftigt sich doch schon bald seit einem halben Jahrhundert mit Klee: als Sammler, als Wissenschafter, als Dramatiker. Kann das nie genug sein?
D: Wahrscheinlich nicht. Paul Klee ist der Kleinmeister unter den Großen der Moderne. Das gilt aber nur für seine Formate. Von seinem geistigen Horizont her betrachtet, ist er ein gewaltiger Kontinent, in dem es immer noch ganze Länder und Schluchten, Gebirgszüge, Seen und Flüsse zu entdecken gibt. Ich verstehe, dass Djerassi von Klee nicht genug kriegen kann. Im Übrigen hat er seine beiden Klee Sammlungen ja aus der Hand gegeben und Museen geschenkt. Seine Klee-Aneignung ist jetzt die eines Kunsthistorikers und Intellektuellen.
K: Mag schon sein. Aber warum schreibt er dann keine Klee Monografie, wieso publiziert er nicht wissenschaftliche Aufsätze über Paul Klee? Warum produziert er nicht wissenschaftliche Sekundärliteratur? Muss es denn die dekadente Mischform des Doku-Dramas sein? Wählt Djerassi die direkte Rede, um verständlich zu klingen, oder weicht er dem dürren Anmerkungsapparat nur aus, weil er seine scharf zugespitzten Thesen nicht ausreichend begründen kann?
D: Ich glaube, er will einfach verständlich sein. Er will Wirkung zeitigen. War das nicht schon sein ganzes Leben so? War es ihm nicht immer schon wichtig, eine größtmögliche Wirkung zu zeitigen? Und ist ihm das nicht auch mit einigen seiner Entdeckungen und Erfindungen wie nur wenigen Wissenschaftern im 20. Jahrhundert gelungen?
Außerdem finde ich, dass seine Bildbelege für seine Thesen, seine Quellengenauigkeit und seine detaillierte Recherche es mit jeder peniblen Klee-Forschung aufnimmt.
K: Geschenkt. Worum geht es ihm eigentlich bei diesem Klee, dem Angelus Novus? Um Klee als Künstler? Oder dreht sich wieder einmal alles nur um den Autor selbst? Djerassi soll ja ziemlich eitel sein.

D: Ich glaube, beides interessiert ihn. Der Künstler Klee und der Interpret Djerassi. Bei diesem ganzen Stück „4 Juden auf dem Parnass" hat man das Gefühl, eine dramatisierte Fassung einer Selbstreflexion, wie sie die besten Tagebücher auszeichnet, in Händen zu halten.
Sicher geht es in dem Stück um vier Juden, und keine kleinen. Aber was sie ausmacht und als Jude definiert, teilen sie bis ins Innerste mit dem Autor des Stücks, mit dem Chemiker Carl Djerassi, dem Emigranten. Und zugleich handelt das ganze Stück davon, was jemanden berühmt macht, und warum. Wann jemand zum Kanon der Musik, Literatur, Philosophie, der Geschichte, der Menschheit gehört. Das interessiert Djerassi. Er möchte auch zum Kanon gehören. Er möchte auf dem Parnass leben, der Nachwelt in Erinnerung bleiben.
K: Das tut er doch längst. Als Erfinder der Antibabypille.
D: Also kommen wir auf Djerassi zurück. Es geht ihm ja in diesem Stück nicht um den ganzen Klee. Genau genommen geht es ihm nur um ein einziges Werk. Vordergründig geht es ihm nur um die Frage, warum diese Zeichnung des Angelus Novus so berühmt wurde, obwohl sie es vom künstlerischen Standpunkt nicht verdiente. Aber durch eine List der Geschichte wurde die Zeichnung zu einem Sammelbecken von Fehleinschätzungen. Djerassi erzählt uns von davon: von diesem Abgrund. Von der in einer Zeichnung sedimentierten Liebe des Opfers zu seinem Henker. Ob das bewusst oder unbewusst geschehen ist, das macht für den im Wien der Psychoanalyse Aufgewachsenen gar keinen großen Unterschied.
K: Sie meinen, es geht um mehr als um die Frage, wann das Werk eines Künstlers zu einem berühmten Werk der Kunstgeschichte wird.
D: Natürlich, das wäre ihm viel zu simpel. Obwohl es schon eine eigene Pointe ist, dass dieses Blatt von 1920 nur deshalb berühmt wurde, weil es einen berühmten Interpreten gefunden hat und der seine – gelinde gesagt – metaphysische Geschichtsthese mit diesem Werk illustriert hat. Übrigens überzeugt mich völlig, wie Djerassi Benjamins Beschreibung dieses Klee brutal demontiert.
K: Sie meinen Schönberg. Schönberg, der rationalste unter allen Komponisten, demontiert die willkürliche Beschreibung Benjamins.
D: Aber ich sagte doch bereits: das ganze Stück ist nichts als eine versteckte Autobiografie, ein Tagebuch voller Selbstreflexionen und Introspektionen. Arnold Schönberg ist Carl Djerassi. Analytisch und kritisch bis zum Gefrierpunkt, unduldsam und autoritär wie der große Meister der Wiener Schule – und rechthaberisch! Damit es niemandem auffällt, kleidet Schönberg/Djerassi seine Rechthaberei in Fragen, die den Antwortenden in eine Sackgasse treiben, aus der es keinen anderen Ausweg gibt, als die Wahrheit. Jeder dergestalt Gefragte

wünscht sich lieber ein Diktat, als so gefragt zu werden. Gegen Schönbergs Fragen ist ein Verhör ein Kinderspiel.

K: Da ist was dran. Aber es ist immerhin eine gewaltige Leistung, dass es dem Autor gelungen ist, nicht nur die Provenienz dieses Angelus Novus von Paul Klee bis in die letzte Verästelung hinein lückenlos schließen zu können, sondern auch noch Benjamins Sekundärausbeutung des Angelus Novus als anmaßenden Willkürakt eines Interpreten bloßzustellen.

D: Ja, das ist schon eine wissenschaftliche Glanzleistung, dem Bild aus dem Besitz von Benjamin über das Versteck in der Bibliothèque nationale in Paris nach New York zu folgen, ehe es über Benjamins Sohn offensichtlich nicht ganz rechtmäßig an Adorno gelangt ist und die Arbeit schließlich durch den letzten Eigentümer, Gershom Sholem, dem Israel Museum geschenkt wurde.

K: Ich gebe zu, so eine penible Provenienzforschung setzt den Geist eines Wissenschafters voraus, nicht die Freiheit des Dichters. Einfühlsamkeit und Neugierde gehen hier eine Symbiose ein, wie sie ein anderer Historiker mit einer poetischen Pranke besessen hat: Golo Mann.

D: Das haben jetzt Sie gesagt, ich würde nie so weit gehen, Djerassi mit Golo Mann zu vergleichen, eher mit Dürrenmatt. Schon, um der Ironie, die beiden eigen ist.

K: Nein, für Dürrenmatt ist Djerassi zu wenig freizügig im Umgang mit Fakten. Außerdem fehlt in seinen Stücken immer der Paukenschlag, die völlige Überraschung. Ihm geht es eher um das aus dem Wissen resultierende befreiende Gelächter der Aufklärung, das Götter entzaubert.

D: Genau das ist es. Entzauberung und Enthüllung: bis der Kaiser ganz nackt ist. Am Glanz kratzen, bis es Katzengold wird. Djerassi liebt die List der Geschichte, den fruchtbaren Irrtum und das Durchschauen falscher Größe. Nichts amüsiert ihn mehr, als wenn er in einem Meisterwerk antiker Skulptur eine neuzeitliche Kopie einer römischen Kopie nach einem griechischen Original entdeckt. Wobei ihn nicht so sehr die richtige Datierung fasziniert. Djerassi fasziniert, wie unser scheinbares Wissen, unsere Vorurteile unser Sehen bestimmen. Das gilt für den Jüngling vom Magdalensberg, den man Jahrhunderte als antikes Stück bewundert hat. Das gilt aber auch für die Anbetung einer Klee-Zeichnung, auf der man vermeintlich sieht, was auf ihr nicht dargestellt ist. Das Wissen, auch das vermeintliche Wissen, bestimmt unser Sein und Sehen. Der Aufklärer Djerassi weist darauf hin.

K: Sie meinen, die eigentliche Pointe besteht für den Autor der 4 Juden weniger darin, dass Klees Angelus Novus durch Benjamins Deutung nachträglich kanonisiert wurde, als dass der Engel möglicherweise Adolf Hitler darstellt.

D: Selbstverständlich. Nur deshalb interessiert sich Djerassi für diese durchschnittliche Arbeit des großen Klee. Das Kapitel in den 4 Juden auf dem Parnass handelt von falschen Versprechungen, falschen Erwartungen und falschen Projektionen. Zur tragischen Komödie wird dieser Budenzauber aber erst dadurch, dass das Bild Hitlers, ein verstecktes Portrait dieses grässlichen Diktators von Benjamin, Adorno und Sholem geliebt und aufgrund seines Tiefsinns bewundert wurde und zum Zeugen des Engels der Geschichte gemacht wurde, der vor jenen Trümmern zurückweicht, die dieser Nazi-Scherge in jenem schlimmsten aller Jahrhunderte überhaupt erst aufgehäuft hat.

K: Sie meinen, Djerassi ist „Satiriker"?

D: Ja! Und zwar einer der kompromisslosesten, den die Geschichte kennt. Alle großen Aufklärer waren Satiriker. Von Petronius über Sebastian Brant bis zu Laurence Stern und Oscar Wilde. Alle großen Aufklärer waren große Demolierer. Sie haben Verbohrtheit und Vorurteile zerstört. Und wenn schließlich kein Stein mehr auf dem anderen bleibt, dann brechen diese Dichter und Denker über den Zustand der Welt in homerisches Gelächter aus.

K: Das ist ja erschreckend. So sehen Sie Djerassi, den Humanisten?

D: Ja! Zuerst sieht er aus der Vogelperspektive zu, wie ganz langsam, Schritt um Schritt, die Deutung Benjamins von Klees Angelus Novus als Engel der Geschichte in sich zusammenbricht. Dann sieht er zu, wie sich dieser Engel ganz langsam, Schritt um Schritt, in den großen Dämon verwandelt. Und dann stößt er wie ein Raubvogel zu: so ein Abbild habt Ihr besessen und als Beleg Eurer Thesen genommen. Und Ihr wundert Euch, dass man in Euch Juden sieht, wo keine sind, weil die Assimilationsleistung alles Jüdische weggeräumt hat? Dass man in Euch Juden sieht, obwohl Ihr nie Juden ward, wie der als Paul Zion Klee denunzierte deutsche Künstler. Wer macht Euch zu Juden? Machen Euch andere zu Juden? Oder macht Ihr Euch selbst dazu? Machen wir uns selbst dazu? Kann man dem Jude-Sein überhaupt entkommen, durch Assimilation oder Flucht in den Protestantismus? Fragen über Fragen. Lauter Fragen und keine Antworten. Auch das zeichnet den Aufklärer Djerassi aus: Der Weg der Erkenntnis führt über Fragen zu weiteren Fragen.

K: Das kennen wir doch seit Sokrates und den Platonischen Dialogen. Sie wollen doch jetzt nicht Sokrates oder gar Platon zu einem Satiriker machen? Auch wenn der fragende Philosoph wie ein Satyr ausgesehen hat, so war er doch der Inbegriff des Humanisten, der etwas schafft, nicht zerstört.

D: Nach den beiden Kriegen, die wir die Großen nennen, ist es schwer geworden, konstruktiv zu bleiben, ohne verlogen zu sein. Auch die vier Juden, die sich Djerassi gewählt hat, waren Großmeister der produktiven Zerstörung: von der Kaballa, die Sholem zerlegt hat, über die süßen Melodien, die Schönberg zerschnitten hat, bis zum Kunstwerk, das Benjamin so lange geputzt hat, bis keine Aura mehr dran blieb sowie dem großen philosophischen System, gegen das Adorno angeschrieben hat.

K: Djerassi ist also ein produktiver Zerstörer? Oder ist er ein destruktiver Produzent? Kann man das eine vom anderen überhaupt trennen, wenn man die Welt erkennt, durchschaut und das Leben zu Ende denkt?
D: Es kommt nur auf das Temperament an. Das Temperament und der Charakter entscheiden, ob man wie Heraklit über die Welt weint oder wie Diogenes über sie lacht. Djerassi hat sich für das Lachen entschieden. So hat er Klee gefunden, oder besser gesagt, Paul Klee hat ihn gefunden. Zwei Gleichgesinnte haben einander entdeckt. Zwei Satiriker, zwei Anti-Pathetiker. Auch davon handelt das Stück „4 Juden auf dem Parnass": Klee war kein Pathetiker. Benjamin hat diese zerbrechliche lächerliche Figur des geflügelten Dämons fälschlich mit Pathos aufgeladen und zum visuellen Fundament seines Geschichtsbildes gemacht. Mit diesem Anspruch konnte er nur in die Irre gehen und dem Komiker Klee Gewalt antun. Und wer entdeckt bei Benjamin diesen Fehler, diesen monströsen geschichtsträchtigen Irrtum einer hypertrophen Interpretation? Ausgerechnet Arnold Schönberg, der Pathetischste unter allen Bewohnern des Parnass. Der eitle Schönberg, der sich nicht genug tun konnte, damit zu prahlen, schon zu Lebzeiten auf den Parnass gekommen zu sein. Dem nie etwas misslungen ist, der nur Banausen kennt oder Menschen, die ihn bewundern. Schönberg ist es ausgerechnet, der Benjamin seine fatale Fehleinschätzung des Angelus Novus vorführt. Das Opfer Benjamin hat ein Bild seines zukünftigen Schergen zu seiner Geschichtsikone gemacht.
K: Djerassi kann Schönberg nicht sympathisch finden, das kann ich nicht glauben. Ausgerechnet mit diesem Schönberg identifiziert er sich und fällt über den armen Benjamin her? Ausgerechnet an seinem geliebten Klee beschreibt er den Prozess der falschen Kanonisierung, der Götzenbildnerei? Warum? Warum?
D: Fragen, nichts als Fragen. Lauter Fragen und keine Antworten.

From the Pill to Paul Klee: Translating Carl Djerassi

Ursula-Maria Mössner

Translators are often forgotten. We are something like the Sherpas that help the expedition to get to the top, and like them we are rarely included in the summit photograph. In literary supplements and magazines, reviewers will praise the style and language of an author and quote long passages of his work without referring to the fact that they are talking about, and quoting from, a translation – let alone mention the name of the translator. I am therefore particularly grateful to the editor of this volume for inviting two of Carl Djerassi's translators and thereby acknowledging our contribution. To illustrate some aspects of this work, its problems and pleasures, as well as the rather special relationship between one author and his translator, I would like to talk about my years as Carl Djerassi's German "voice."

The beginning: *Cantor's Dilemma*

In April 1990, I signed the contract to translate into German *Cantor's Dilemma* by Carl Djerassi. A couple of weeks earlier, I had had a phone call from a Swiss publishing house and the first question had been: "Do you know anything about chemistry?" Had I been honest, I would have had to say "No," but since my daughter was at that time an undergraduate student of chemistry I did not hesitate to say: "I do." The caller seemed greatly relieved and right away offered me a novel, set in Academia and dealing with a scientific discovery that is awarded the Nobel Prize. He went on to praise the book, the first novel of a famous American chemist, the "Father of the Pill."

I read the book, liked it and so, quite cheerfully, began to translate. Cheerfully because at that time – I am ashamed to admit it – I had never heard of Carl Djerassi. Only while working on the translation, when I happened to mention the name in academic circles, did it begin to dawn on me with whom I had become involved – and, later still, what I had let myself in for. Some weeks after the translation had been completed and sent to the publisher's, the phone rang and a deep and sonorous voice said: "This is Carl Djerassi."

For the author who, by his own words, had not read a work of fiction in German since being forced to leave Austria in 1938, it must have been a rather ambivalent, maybe even unsettling experience to read the novel he had written in English translated into his native tongue. All the more so since the translation was in contemporary German, as opposed to the German spoken fifty years earlier during Carl Djerassi's childhood and youth in Vienna.

So we began to discuss the translation, arguing about words and phrases, and particularly about Anglicisms that have found their way into everyday German and which the author did not like one bit. Naturally, he had far "better" solutions on hand, and I had a hard time of it to justify certain idiomatic expressions and to suggest cautiously that his German might be slightly old-fashioned and outdated. It was a very long and not always easy phone call, but then Carl Djerassi said a most wonderful sentence: "You are the judge." With these words he generously conceded that I was to have the last word regarding the translation. I would be the one to decide which of his suggestions to accept and which to reject – and I deeply appreciated this kind gesture. Strictly speaking, at that time, his decision was justifiable to some degree, but by now Carl Djerassi's German is so perfect that he could easily write his books in German and make me redundant.

From the Pill to Paul Klee

This was the beginning of a very fruitful and friendly cooperation. In 1991, after *Cantors Dilemma* (for the German title, all we had to do was to leave out the apostrophe) was published in Germany to great acclaim, we met in person and, if I may say so, immediately took to each other. In the course of the next ten years Carl Djerassi entrusted me with the translation of four more novels in the genre of "science-in-fiction," two collections of short stories, three autobiographical works, and some poems. Later, when he began to write science-in-theatre, four plays and two radio plays were added to this list.

An author who is an organic chemist by training, who has done research in industry and taught at universities, who collects art and supports artists, a man who writes novels and plays to acquaint the reader with the fields of science and the humanities, in short, an intellectual polygamist like Carl Djerassi, forces his translator to be prepared to enter – and conquer – unknown territories. But, being a good pedagogue, he fortunately heeds Einstein's dictum: "Things should be made as simple as possible, but not any simpler."

During the first years, particularly while translating *The Pill, Pygmy Chimps, and Degas' Horse* (the German title *Die Mutter der Pille* gives no hint of the variety of subjects that are dealt with in the book), I practically lived with the German standard textbook of organic chemistry, struggling with bonds and chains and the structure of steroids, trying to come to grips with the synthesis of cortisone, the de-aromatization of progesterone and finally with norethindrone and the development of the Pill. Apart from being informed about pygmy chimps and gorillas using sign language, about perestroika, pest control and Pugwash, I learned quite a few useful titbits, like how to buy art, how *not* to make a movie and, most important of all, how to get a cockroach to take the Pill.

In *The Bourbaki Gambit*, I was confronted with the Polymerase Chain Reaction, while in *Menachem's Seed* and *NO* I was flooded by a host of very heterogeneous subjects such as nuclear weapons and disarmament, reproductive medicine and intracytoplasmic sperm injection (ICSI), the founding and IPO of a bio-tech company, nitric oxide and erectile dysfunction, a fictitious 17th century drama supposedly set to music by Händel, plus a touch of Koran, a little Hebrew and a lot of Judaism.

Recurring themes in Carl Djerassi's books have always been art, music, and literature. It began quite harmlessly in *Cantor's Dilemma* with chamber music and Egon Schiele; the charming short story "The Toyota Cantos" called for quotes from Dante; the play *Ego* featured the Portuguese poet Fernando Pessoa and his heteronyms; and in *Four Jews on Parnassus* four great minds, Walter Benjamin, Theodor Adorno, Gershom Scholem, and Arnold Schönberg come together – to talk about Paul Klee who had already popped up in earlier works, not surprising given the fact that he is Carl Djerassi's favorite painter.

Another feature of Carl Djerassi's books and plays is sex. In *Cantor's Dilemma* this consisted of one-night stands and the sucking of toes, while in *Ego* the peeling of a mango provided a delicate foreplay; in *Taboos*, things became decidedly more hard-core, since a central point of this play is the artificial insemination of a lesbian by another lesbian – by means of a turkey baster to boot; and it culminated so far on Parnassus with a long list of sexual practices that made me blush all over. In fact, to be honest, some of them are still a mystery to me.

Naturally, I had neither the extensive knowledge needed to do justice to Djerassi's books, nor could I acquire it at short notice; consequently I needed the help of experts in a variety of fields. And I had no problem finding them – thanks to my author who seems to know everybody and generously put me in contact with exactly the right person. The words "I am the German translator of Carl Djerassi" opened all doors for me. The controller of a pharmaceutical company patiently explained to me the widely differing company structures in Germany and in the United States and took me step by step through the stages of a company's Initial Public Offering; in Berne, Switzerland, an art dealer explained to me in detail specific painting techniques of Paul Klee; and a renowned specialist in reproductive medicine provided me with the newest publications in his field und was always ready to answer questions.

This space of some twenty years in which I translated Djerassi might well be summed up under the title "from the pen to the worldwide web." In 1990, most translators were still using a typewriter and did their research in public libraries, and all publishers still demanded that translations were sent in on paper. Since Carl Djerassi examined every translation punctiliously, our communication took place in the form of parcels and phone calls – between San Francisco and Ulm where not only the different time zones, but also the sheer length of our conversations made for hot ears and stiff arms. Later we would often meet in London, in Carl Djerassi's lovely flat, to discuss my translation, often after an excellent meal served by the author himself.

After some arm-twisting on the part of my author, I bought a fax machine and instead of parcels I now got long faxes. Of course, by that time Carl Djerassi was using a computer, and when I myself had acquired a PC, he was already happily surfing the worldwide web. So I had no choice but to go online myself. That was a great improvement, especially for the author, who, as soon as he started work on a book or play, was now able to pass on any changes, deletions, or additions without loss of time simply by pushing a button – for the translator quite often a mixed blessing. Small wonder then that the publication of the German translation often predated the publication of the American original – which, to my astonishment, occasionally had a different ending!

Having to deal with so many different subjects was not only a great challenge but also a great personal gain. Translating Carl Djerassi gave me the chance to enlarge my own knowledge, particularly in fields that I would otherwise not have come in close contact with, and allowed me to meet a number of distinguished and fascinating people whose acquaintance I could only have made in this context. For all of this I am deeply endebted and grateful.

Challenges

And now some practical examples of how I dealt with specific difficulties offered by puns, anagrams and other generic and stylistic variations that had me puzzling for days or weeks. It is hard to imagine for anybody, other than the translator, how much effort can go into a simple little wordplay.

In *Menachem's Seed,* an Israeli and a Palestinian are talking about nuclear terrorism, including the Syrian nuclear reactor Osirak, which some months later was indeed bombed and destroyed by Israel. The Palestinian ends the conversation with a warning in the form of a Spoonerism:

"Fighting a liar, lighting a fire."

There is no easy translation for such a sentence. One does not only have to find a German equivalent but one that lends itself to a Spoonerism, which can take a lot of thinking and experimenting. But whatever I tried, no luck. After having brooded over the problem for a couple of weeks, I asked the author if he would be willing to dispense with the Spoonerism and accept a suitable free translation. But Carl Djerassi was adamant: it had to be a Spoonerism, period. And a Spoonerism it is, because in my frustration I had a brain wave:

"Versuchte Rache, verruchte Sache."

Here an example from *Cantor's Dilemma*. When first meeting the protagonist, the fabulous Paula Curry wants to know his first name:

Cantor: "People just call me 'I.C.'"
Paula: "Icy? You don't seem icy to me."
Cantor: "Not icy. Eye cee," he enunciated the letters carefully.

My first idea was to use the initials as pronounced in German where they stand for a fast train, and to let Paula say something like: "So schnell sehen Sie gar nicht aus" ("you don't seem so fast to me"). But I discarded the idea. Since the reader would hold a book in his hands and see the initials in their written form, I decided to keep the English pronunciation and simply explain it by using similarly sounding German words.

Cantor: "Ich werde einfach I.C. genannt."
Paula: "Ei, Sie? So wie 'ei, du'?"
Cantor: "Nicht 'ei, Sie!'," sagte er und buchstabierte akribisch seine Initialen.

And now an anagram from the novel *NO*. In a discussion about how to fill a certain position, a scientist answers with an anagram.

Pithecanthropus erectus.
Pursue the person; catch it.

As a rule, I solve anagrams with the help of *Scrabble* pieces, for neither Google nor any other search machine has ever offered me useful suggestions that met the required meaning. So the *Scrabble* pieces were laid out and for days on end the whole family was busy arranging and rearranging them. Now, in a book about nitric oxide and male erection it would have been desirable to preserve the "erectus" in the German translation. But unfortunately we were unable to. In the end we came up with the following solution:

Pithecanthropus erectus.
Ratetip: Sucht euch Person.

Compared with the problems mentioned above, my last example was child's play. In *Phallacy,* an art historian asks a scientist who specializes in the dating of bronzes:

"What do you think of Klee?"
"I'm into bronze not clay."

Here of course it is the fact that in English "Klee" and "clay" are pronounced more or less the same. But this is a play, and therefore the words will not be read, but heard, and since "Klee" means "clover," I came up with:

"Was halten Sie von Klee?"
"Ich bin Chemiker und kein Botaniker."

It does not end on Parnassus

In the course of all these years, counting novels, collections of short stories, poems, autobiographical works, plays, and articles, I translated at least one work of Djerassi's per year – making me indeed something of his court translator. And I have to commend my author for having been "unfaithful" only three times, that is with the plays *An Immaculate Misconception*, *Calculus*, and *Oxygen*. Nevertheless, the two of us managed to smuggle something from me into each one of them, be it a couple of pages, a short paragraph or at least a word play.

And so the author and his translator joined four Jews on Parnassus. The author made the ascent effortlessly and has already found his designated seat up there – as a scientist and as a writer. The translator fared less well; she felt a bit like the wives of the four geniuses: struggling uphill, fighting for breath all the way, catching a glimpse of the hallowed halls and then clambering down to earth again.

But Parnassus is not the end. My author is still going strong and I sincerely hope that he will keep me busy for years to come. In *This Man's Pill*, where he describes how the publication of his fiction in German changed his attitude towards Austria and Germany, Carl Djerassi paid me the greatest compliment possible: "It took a modern German voice, and a woman's voice at that, to bring me to terms with my European origins."

Bibliography of works by Carl Djerassi translated by Ursula-Maria Mössner

Cantors Dilemma. (*Cantor's Dilemma*). Zürich: Haffmans, 1991.
Der Futurist und andere Geschichten. (*The Futurist and Other Stories*). Zürich: Haffmans, 1991.
Die Mutter der Pille. (*The Pill, Pygmy Chimps, and Degas' Horse*). Zürich, Haffmans, 1992.

"Die Uhr läuft rückwärts." ("The Clock Runs Backward.") *Der Rabe* 35. Zürich 1993. 94-96.
Das Bourbaki Gambit. (*The Bourbaki Gambit.*) Zürich: Haffmans, 1993.
Marx, verschieden. (*Marx, Deceased.*) Zürich: Haffmans, 1994.
Menachems Same. (*Menachem's Seed.*). Zürich: Haffmans, 1996.
Von der Pille zum PC (*From the Pill to the Pen*). Zürich: Haffmans, 1998.
NO (*NO*). Zürich: Haffmans, 1998.
Wie ich Coca-Cola schlug. (*How I Beat Coca-Cola*). Zürich: Haffmans, 2000.
This Man's Pill – Sex, die Kunst und Unsterblichkeit (*This Man's Pill.*) Innsbruck: Haymon, 2001.
Stammesgeheimnisse – Zwei Romane aus der Welt der Wissenschaft. Innsbruck: Haymon, 2002.
Ego (Roman und Theaterstück). Innsbruck: Haymon, 2004.
Ego (Hörspiel). Westdeutscher Rundfunk Köln, 2004.
Aufgedeckte Geheimnisse – Zwei Romane aus der Welt der Wissenschaft. Innsbruck: Haymon, 2005.
Phallstricke (Hörspiel). Westdeutscher Rundfunk Köln, 2005.
Phallstricke/Tabus – Zwei Theaterstücke aus den Welten der Naturwissenschaft und Kunst (*Phallacy/Taboos*). Innsbruck: Haymon, 2006.
Vier Juden auf dem Parnass (*Four Jews on Parnassus*). Innsbruck: Haymon, 2008.
Vorspiel. Theaterstück. Innsbruck und Wien: Haymon 2011.

Parnassus

Four Jews on Parnassus:
Carl Djerassi's Refiguration of the German-Jewish Parnassus

Erin McGlothlin

In *Four Jews on Parnassus: A Conversation*, Carl Djerassi achieves in fiction what would have been impossible in real life: he brings together in one space and at one time four of the most prominent intellectuals and artists of the twentieth century, three of whom were well acquainted with each other and one of whom had never met the other three, and puts them into intensive direct dialogue with each other. Although the five staged conversations between the men in *Four Jews on Parnassus*, which Djerassi calls "docu-dramatic scenes" (xvi), are based on meticulous research and documentation (as Djerassi's impressive bibliography and archival notes demonstrate) and often derive from actual statements the men made in their writing or in interviews, the encounter itself is wholly imagined, for at no time did all four historical figures participate in a common dialogue. Djerassi makes this mostly authentic and at the same time fully fictional four-way conversation between historical figures possible not by rewriting their life history so that the encounter be rendered feasible, but by locating them in a sort of illustrious canonized afterlife on the site of Parnassus, the mountain that in ancient Greece was home to Apollo and the muses. As Djerassi tells us at the beginning of the first dialogue, "Parnassus is a commonly accepted metaphor for the ultimate recognition of literary, musical, or intellectual achievement. Arrival on this exalted peak demonstrates that the process of canonization is complete." (1) By adopting the ancient metaphorical space of Parnassus and refashioning it into a concrete setting for his staged encounter, Djerassi is able to produce this otherwise impossible dialogue.

Djerassi's choice of setting is particularly important because Parnassus has a long and rich cultural history not only in ancient Greece, where it was one of a number of mythical sites in a widespread sacred geography, but particularly in Germany and Austria, where the ancient site was appropriated as a trope for the very idea of an exalted German-language literary canon. Parnassus's afterlife in

German culture is especially relevant given the national identities of Djerassi's four intellectuals, one of whom was Austrian, the others of whom possessed, at least originally, German citizenship. At the same time, however, as the title of Djerassi's dialogue indicates, his protagonists – Theodor W. Adorno, Walter Benjamin, Gershom Scholem and Arnold Schönberg – are also Jewish, which necessarily complicates their respective relationships to German and Austrian culture and the German language. As the four men discuss in the five chapters of the dialogue, each of which centers in turn on the particular themes of death, love and sex, art, Jewish identity and intellectual and artistic legacy, their relationship to German culture as Jews was a critical factor not only in their intellectual and artistic work and in the German and Austrian reception of that work but also in the posthumous renown that sustains their current residence on Parnassus. As Djerassi explains in his Preface:

Why did I pick this particular foursome? Because all four belonged to the peculiar subset of German and Austrian bourgeois Jews of the pre-World War II generation who often were more Berlinish or Viennese than their non-Jewish compatriots. None was deeply religious; some of them were essentially secular. This is also the generation and social subset to which I belong, and my own personal experience with the indelible effects of growing up as a secular Jew in Vienna in the 1930s made me want to examine the range of the meaning of *Jew* through four individuals who responded so differently to that label. (xv)

Djerassi's interest thus lies in how these four German-Jewish and German-Austrian figures on Parnassus negotiate each side of their hyphenated identities and how each navigated their professional and personal lives through the catastrophic historical events that affected European Jews in the twentieth century. Moreover, as Djerassi admits above, his imagination of the ways in which Benjamin, Adorno, Scholem and Schönberg perceive their Jewish identity from their posthumous perch on Parnassus is related to his autobiographical interest in his own Jewish identity: "But there is more to my choice of these four European Jews: I recognized themes in their lives that I also wanted to examine in my own as I approach its end." (xvi) Djerassi thus links his investigation of the four intellectuals to his autobiographical impulse and to questions about his own Jewish identity as he himself approaches Parnassus as a result of his work as both a pioneering, award-winning scientist and an innovative writer of literary texts. In this way, although his text clearly focuses on the lives and intellectual legacies of the four men, it bears, like many of Djerassi's texts, traces of autobiographical resonance.

Djerassi's choice of setting for the dialogue between the four great men is the mountain of Parnassus, which he refigures into a site that is ideally suited for the posthumous review of one's life. Why exactly does he choose Parnassus for his staged encounter between the four men? On the one hand, it seems the ideal solution to the problem of representing Benjamin and company in their posthumous state, for it allows him to imagine their mutual interaction after their deaths as a sort of meta-reflection on their lives without concurrently positing a sort of afterlife for them. Judaism is somewhat vague when it comes to any specific notion of an afterlife, so it is only fitting for both author and characters, in keeping with their secularity, that Djerassi would avoid creating a religious afterlife and instead would invent an intellectual one. One might thus consider Djerassi's Parnassus the secular, acculturated counterpart to Mount Sinai, a site of Jewish cultural identity as opposed to religious affiliation to Judaism. Moreover, the act of placing Benjamin, Adorno, Scholem and Schönberg in a sort of heaven would run counter to Djerassi's explicit aim in his biographical view of them, which focuses on presenting a "humanizing" view of their lives and work; as he writes, "hagiography is an outcome I have tried to subvert throughout this work." (180) On the other hand, the motif of Parnassus carries with it a long cultural history, especially in German, as I mentioned earlier. By looking more closely at the trope of Parnassus in German discourse from the early modern period to the early twenty-first century, we are able to understand more fully not only its cultural import, particularly with regard to the relationship between the German and the Jewish in German literature and culture, but also the implications of Djerassi's choice to refigure it for his present purposes.

Djerassi gives a helpful – albeit brief – encyclopedia definition of Parnassus at the beginning of the first chapter:

Par·nas·sus (pär-năs'əs) also Par·nas·sós (-nä-sôs'): A mountain, about 2,458 m high, in central Greece north of the Gulf of Corinth. In ancient times it was sacred to Apollo, Dionysus, and the Muses. The Delphic Oracle was at the foot of the mountain. Metaphorically, the name "Parnassus" in literature typically refers to its distinction as the home of poetry, literature, and learning. (1)

Djerassi's short description is echoed in the encyclopedic literature; contemporary reference works that list "Parnassus" – even those dedicated specifically to Greek myth and literature – relate basically the same scant information. This absence of extended analysis of the import of Parnassus indicates that, in classical studies, it is considered just one site in a rich, multiple and varied mythical landscape; for this reason it does not really stand out as something extraordinary or as a locale that deserves special analysis. It is only in its post-classical manifestations that the concept of Parnassus takes on special import apart from the rest of Greek mythological geography.

Post-classical appropriations of Parnassus have occurred above all in German literature, where the trope appears as a specifically German motif already in the early modern period. Johann Rist, one of the most important German Protestant poets of the seventeenth century, published a collection of poems in 1652 entitled *Neuer Teutscher Parnass* [*New German Parnassus*]. In his preface, Rist explains at length the geographical and cultural significance of Parnassus and why he chooses this "heathen" site as the motif for his poems:

> In such a way one should behold and observe Mount Parnassus in Greece, of which so many wonderful things have been written, sung and versified; for this reason I also want to allude in this preface to Parnassus in the dominion of Phocis and its character with regard to a number of aspects. Yes, says the careful reader at this point. I will easily concede all of this. I can certainly believe that such a mountain in Greece existed and is probably still there. But why must your present book also be called Parnassus, like the heathen mountain there in Greece? You could have given it the name of a mountain in Christendom or even another title .... Our Rist has his own Parnassus in Germany; it therefore follows that he must be the German Apollo and thus the most prominent of all poets.[1]

As Rist further explains, he decides to locate a new, alternative Parnassus in Germany not because he wishes to claim that he is the German Apollo, but rather because the landscape in Holstein in which he composed his poetry (which he calls "my German Parnassus in Holstein"[2]) provided him with the same sort of inspiration experienced by Apollo on the Greek Parnassus. With this gesture of relocating and renaming Parnassus, Rist thereby elevates German space as a site of poetic production and at the same time refigures Parnassus as a site of, in particular, German poetry.

It is in the eighteenth century, however, when the German appropriation of the concept of Parnassus becomes more widespread. Not surprisingly, at a time of both the advent of German classicism and the rising importance of the artist and the poet, Parnassus provided a malleable motif that could be reconstituted for the art of the day and reshaped to suit a variety of aesthetic aims. Johann Joachim Winckelmann, known as one of the greatest proponents of German neoclassicism and the founder of modern art history, had a profound influence on art in the second half of the eighteenth century; he helped shift prevailing artistic taste

---

[1] "Auff eine solche Ahrt nun sol man auch den in Griechenland gelegenen Berg Parnass, von welchem so viel wunderbares Dinges ist geschrieben, gesungen und gedichtet, ansehen und betrachten. Darum auch Ich in diesem Vorberichte dieses in der Phoeischen Herrschaft gelegenen Parnassus und seiner Beschaffenheit mit Mehrerein habe erwähnen wollen. Ja, spricht man nun hierauff der sorgfältige Leser: Ich gebe dieses alles gerne nach, Ich kan endlich wol glauben, daß ein solcher Berg in Griechenlande gewesen auch noch wol daselbst sei zu finden. Warum aber muß eben dein gegenwertiges Buch auch Parnassus heissen, wie dort der heidnische Berg in Griechenland? Du hettest Ihm ja wol den Namen eines in der Christenheit gelegenen Berges, oder auch wol gahr einen anderen Titul können geben .... Hat unser Rist schon einen eigenen Parnass in Teutschland. So muß ja folgen, daß Er auch der teutsche Apollo und also das Haubt aller Dichter sei" (from pages 7 and 14 of the "Nohtwendiger Vorbericht"). All translations are my own, E.M.

[2] "Mein Teutscher Holsteinischer Parnass" (from page 1 of the "Nohtwendiger Vorbericht").

from Roman art to Greek art and scholarly attention from historical referents of ancient art to its intrinsic stylistic qualities. Winckelmann's role in the German neo-classical appropriation and redefinition of classical motifs – especially that of Parnassus – is considerable. Under his influence, Anton Raphael Mengs, who was widely regarded as the greatest living artist of his day, painted the 1761 ceiling fresco for the Villa Albani in Rome entitled "Parnassus," which depicts Apollo and the Muses. This painting, which reworks the composition of Raphael's "Parnassus" at the Vatican Palace, was hailed as one of the most important examples of neoclassicism and helped to secure the Parnassus motif in the German neo-classical imagination.

Another prominent figure who reappropriates Parnassus at the end of the eighteenth century is Christoph Martin Wieland, an important writer of the particularly German literary school of neo-classicism, namely Weimar classicism. Wieland assumes the image of Parnassus for his column in the literary journal *Der neue teutsche Merkur* [*The New German Mercury*] which he titled "Der Teutsche Parnaß" ["The German Parnassus"], thus once again refiguring Parnassian inspiration for German literary production. In Wieland's imagination, Parnassus is the seat not only of poetry, but of German poetry and of the German poet in particular. His Parnassus is above all the exclusive residence of the genius poet, who, sequestered from the mundane on his lofty peak, pursues his higher calling.

One cannot make any sort of binding claim about the history of German literature without reference to one of these genius poets in particular, Johann Wolfgang von Goethe, and with respect to the motif of the Parnassus, Goethe does not disappoint. He, too, refigures Parnassus for German poetry; he even includes a brief reference to it in his masterwork *Faust I*. His most extended representation of the mythical site of poetry can be found in his 1798 poem "Deutscher Parnaß" ["German Parnassus"], originally entitled "Sängerwürde." The poetic "I" of this poem is the custodian of Parnassus, who since his boyhood has observed with wonder the poetic activities of the muses. Midway through the poem, the poetic voice watches with horror as the holy Parnassus is invaded by a "verwegendes Geschlecht" ["foolhardy race"], who proceeds to destroy its delicate beauty: "Ah, oe'r every plant they rush! / Ah, their cruel footsteps crush / All the flowers that fill their path! / Who will dare to stem their wrath?"[3] (219) He calls for the invaders to be repelled: "From our boundaries haste away, / From the god's dread anger fly! / Cleanse once more the holy place, / Turn the savage train aside! / Earth contains upon its face / Many a spot unsanctified; / Here we only prized the good. / Stars unsullied round us burn."[4] (221) In Goethe's satirical poem, the quiet sanctuary of

---

[3] "Ach, die Büsche sind geknickt! / Ach, die Blumen sind erstickt! / Von den Sohlen dieser Brut, / Wer begegnet ihrer Wut?" (716)

[4] "Fliehet vor des Gottes Grimme, / Eilt aus unsrer Grenze fort! / Daß sie wieder heilig werde,

Parnassus, the seat of the most elevated art of poetry, is trespassed upon by profane hordes that besmirch the sacred space and pollute the pure tones of its song: "Fain I'd think myself deluded / In the saddening sounds I hear; / From the holy glades secluded / Hateful tones assail the ear."[5] (220)

In Goethe's representation of Parnassus – and again, we have here not the Greek Parnassus, but an explicitly German one – the lofty residence of the literary genius is thus overrun by mediocre poets who, scream and shriek as they might, neither possess the requisite genius and talent to create worthy literature nor the respect for the higher moral calling of the poet. Goethe thus ironically critiques that which Parnassus had become by the late eighteenth century – a site of hagiolatry and worship of the poet's moral stature. As Daniel Jacoby, the German-Jewish Germanist, writes in his 1893 analysis of the poem, this critique goes hand in hand with Goethe's rejection of the notion of the writer's morally elevated stature: "In both his youth and old age Goethe stressed that the art of poetry should impartially represent human striving and suffering and not preach morality, an idea he, as a poet, proved time and again without denying the cultivated influence of art."[6] In "Deutscher Parnaß" (which in many ways resembles Heinrich Heine's later poem "Die Wanderratten" ["The Wander Rats"]), in which the gates to Parnassus must be heavily guarded to keep the space of German poetry holy and pure, Goethe parodies the notion of a calcified, moralistic literature and implicitly advocates for a notion of literature that, rather than separating itself apart from life, engages actively with it. In this regard, at least, there is a connection between Goethe's depiction of the literary Parnassus and Djerassi's later imagination of it. By explicitly refusing to make his own figuration of Parnassus into a site of the monological worship of Adorno, Benjamin, Scholem and Schönberg ("hagiography is an outcome I have tried to subvert throughout this work"), Djerassi, perhaps unconsciously, endorses the Goethean critique.

Despite Goethe's satirical critique of the ways in which the German Parnassus – and the world of artistic production it represented – had become an exclusive domain of poetic genius (a domain in which Goethe himself reigned supreme; as Bayard Taylor wrote in 1895, Goethe is "the central figure of the great age of German literature – the god, he might be called, who sits alone on the summit of the German Parnassus", 304), during the nineteenth century the notion of the Parnassus as the site of German literary brilliance reified during a time in which

---

/ Lenkt hinweg den wilden Zug. / Viele Boden hat die Erde / Und unheiligen genug. / Uns umleuchten reine Sterne, / Hier nur hat das Edle Wert." (718)

[5] "O wie möcht ich gern mich täuschen; / Aber Schmerzen fühlt das Ohr; / Aus den keuschen, / Heiligen Schatten / Dringt verhaßter Ton hervor." (717)

[6] "Daß die Dichtkunst das Thun und Leiden des Menschen unbefangen darzustellen hat, nicht Sittlichkeit predigen soll, hat Goethe in der Jugend wie als Mann und Greis betont und als Dichter immer bewährt, ohne damit den veredelnden Einfluss der Kunst zu leugnen." (204-205)

the German national literary canon itself solidified in the absence of a German nation. The Parnassus became a signifier of this construction of literary greatness.

In a mostly unknown play from 1820 called *Der Deutsche Parnass* [*The German Parnassus*], written by Adolf Wilhelm Schneider under the pseudonym Dichterecht Ehrendeutsch, the luminaries of the German canon, such as Opitz, Gryphius, Klopstock, Herder, Lessing, Wieland and Schiller, reside on the German Parnassus. Goethe, who is not yet dead, is also there; he warns the poets of a possible danger to "Germany's honor" with the "countless bands of German bards who vexed you so when you were alive; anyone who sang a little verse, rhymed or unrhymed, anyone who glued together words; they're all coming to extort Parnassus."[7] Goethe then develops a sort of entrance exam for all poets who wish to enter the sacred space: "Let us protect the German name and rescue it from a death of shame ... Whoever comes to gain entrance here has to pass a poetic test." The minor writer Christoph August Tiedge fails and is turned away; Friedrich Schlegel and August Wilhelm Schlegel, on the other hand, pass with flying colors and are admitted to Parnassus. In Dichterecht Ehrendeutsch's figuration of Parnassus, the motif of the sacred mountain as an exclusive site of German genius is stripped of a Goethean satirical critique and is instead presented without irony as the pinnacle of poetic production (an assertion that is itself performatively ironic, given the childish, awkward verse in which Dichterecht Ehrendeutsch writes). At the same time, in this iteration, Parnassus is at its most German, providing the foundation for an absent national identity. Such focus on the exclusivity of the pure realm of Parnassus that is necessarily and intrinsically German will later be used to justify anti-Semitic notions of who belongs on the German Parnassus and who does not.

In the latter part of the nineteenth century, the motif of the Parnassus is once again used to define the exclusive sphere of German literature. In 1861, Johannes Minckwitz, the Leipzig professor and author, compiled the 900-page literary encyclopedia *Der neuhochdeutsche Parnaß, 1740 bis 1860: Eine Grundlage zum besseren Verständnisse unserer Litteraturgeschichte* [*The New High German Parnassus, 1740 to 1860: A Foundation for a Better Understanding of Our Literary History*], which attempted to review the German literary canon of the late eighteenth and early nineteenth centuries. In its introduction, Minckwitz writes:

A sublime joy awakens the contemplation of the beginnings and advances that German poetry has made within the period of approximately twelve decades. There has emerged a succession of poets, some of whom are of the greatest talent, others of whom are even more exquisite; the former have scaled the highest peaks of our Parnassus, the latter have vaulted its shining pinnacle from which they have flung down to the nation abundantly

---

[7] ' Der deutschen Sänger ungezählte Schaaren, / Die euch im Leben oft so lästig waren, / Wer je ein Verslein sang, gereimt / Und ungereimt, wer je mit Worten geleimt; / Sie kommen All', den Parnass zu erzwingen.' (71)

blooming wreaths. Among those so many a fragrant with evergreen, so that we can justifiably say with pride: our literature is at least on par with regard to intellectual quality and content with every other European literature.[8]

Parnassus is thus once again figured in terms of the German national identity; it is a privileged site of German creativity where an elite corps of German poets reigns over the German literary landscape. In this example of the cultural imagination, Parnassus is appropriated to create a mythical concept of a nation located in literature.

In 1912, the German-Jewish journalist Moritz Goldstein takes up the motif of Parnassus in an influential article in the journal *Der Kunstwart* entitled "Deutschjüdischer Parnass" ["German-Jewish Parnassus"] that has since become a key text for the study of German-Jewish cultural history. The title of the article is deceptive; rather than announcing a new figuration of the existing concept of the German Parnassus that, fifty years after Minckwitz's nationalistic definition, expands the canon of German literature and redefines German genius to include the work of German-Jewish writers who had left the ghetto and entered the German intellectual sphere over a century earlier, Goldstein's article evokes the German-Jewish Parnassus only to then announce its impossibility. In fact, whereas the "German Parnassus" is evoked a couple of times in the article, nowhere in the main text does Goldstein mention a German-Jewish Parnassus; it exists as hypothetically only in the title.

Goldstein's main goal in the essay is to openly question the status of Jewish participation in German culture. As he argues, Jews have entered German cultural life – meaning the press, the theater and literary life – in unprecedented numbers, showing an extraordinary passion for and commitment to German culture. He mentions in this vein particularly Heine, who imagined his own Parnassus in his poem "Der Apollogott" ["The God Apollo"], in which what a star-struck nun takes for Apollo turns out to be a failed rabbi. However, Goldstein believes this passion to be one-sided; he argues that German-Jews, believing that they are taking part in a two-way cultural exchange with their non-Jewish German counterparts, actually contribute to a culture that not only continues to exclude them, but also minimizes the importance of their achievements:

---

[8] "Eine erhabene Freude weckt die Betrachtung der Anfänge und Fortschritte, welche die deutsche Poesie innerhalb eines Zeitraumes von ungefähr zwölf Jahrzehnten gemacht hat. Hervorgetreten sind eine Reihe theils der größten Talente, theils solcher, die zu den vorzüglicheren gehören; die Einen erklommen die höchsten Spitzen unseres Parnasses, die Andern schwangen sich zu irgend einer glänzenden Zinne desselben auf, von welcher sie der Nation blüthenreiche Kränze herabwarfen. Unter ihnen duften so viele immergrüne, daß wir mit Stolz zu sagen berechtigt sind: unsere Litteratur steht jeder andern europäischen am geistigem Werthe und Gehalte mindestens gleich." (iii)

The Jews have taken on the Germans' mission as their own; more and more it appears as if German cultural life were passing into Jewish hands. But the Christians neither expected nor wanted this when they granted the pariahs in their midst participation in European culture. They began to resist, they began to call us foreign again, they began to consider us a danger in the temple of their culture. And so now stand before the problem: We, the Jews, administer the intellectual property of a people that denies us the qualification and the capacity for doing so.[9]

Goldstein locates what he sees as a failed cultural exchange in the impossible concept of a "German-Jewish Parnassus," an imaginary site of an equally imaginary German-Jewish cultural reciprocity. He conceives his Parnassus in a very different way than do the authors of the previous manifestations we have reviewed; for him, Parnassus is neither the site of poetic inspiration, as it is for Rist, nor the resting place of genius, as in the eighteenth century, nor the positive symbol of national literary identity it becomes in the nineteenth century. At the same time, however, one can see elements of these previous evocations of Parnassus in Goldstein's conception. Like the poetic "I" in Goethe's poem, German Jews are figured as the custodians of German culture; however, whereas Goethe's poetic subject is a part of the elevated poetic domain he is charged with watching over, German Jews, in Goldstein's view of Parnassus, are largely excluded from the cultural sphere they so enthusiastically safeguard. Figured in the terms of Dichterecht Ehrendeutsch's figuration of Parnassus, Goldstein's German Jews are those against whom the German Parnassus and "Germany's honor" must be defended; like the man from the country in Kafka's "Vor dem Gesetz" ["Before the Law"], they stand outside the gates of the Parnassus that guard the home of the literary figures they so passionately revere, waiting patiently but ultimately futilely for entry. With this vision of a German Parnassus barred against Germany's Jews, Goldstein underscores the mythical character of a joint German-Jewish cultural project. In his essay, the Parnassus thus becomes a signifier for the impossibility of the German-Jewish endeavor, which in his view is not only unattainable but has also been doomed to failure since its very inception.

The critical thrust of Goldstein's essay was taken up in a debate about the status of Jews in German and Austrian culture that was prominent during the first decades of the twentieth century with figures such as Theodor Herzl and Walter Benjamin and again after the Holocaust by Jewish intellectuals such as Gershom

---

[9] "Die Aufgaben der Deutschen haben die Juden zu ihrer eignen Aufgabe gemacht; immer mehr gewinnt es den Anschein, als sollte das deutsche Kulturleben in jüdische Hände übergehen. Das aber hatten die Christen, als sie den Parias in ihrer Mitte einen Anteil an der europäischen Kultur gewährten, nicht erwartet und nicht gewollt. Sie begannen sich zu wehren, sie begannen wieder uns fremd zu nennen, sie begannen, uns im Tempel ihrer Kultur als eine Gefahr zu betrachten. Und so stehen wir denn jetzt vor dem Problem: Wir Juden verwalten den geistigen Besitz eines Volkes, das uns die Berechtigung und die Fähigkeit dazu abspricht." (283)

Scholem and Dan Diner. Scholem, following Goldstein, objected vehemently to what he saw as the "myth of the German-Jewish dialogue." In an open letter from 1962, he writes:

I deny that there has ever been such a German-Jewish dialogue in any genuine sense whatsoever, i.e., *as a historical phenomenon*. It takes two to have a dialogue, who listen to each other, who are prepared to perceive the other as what he is and represents and to respond to him. Nothing can be more misleading than to apply such a concept to the discussions between Germans and Jews during the last 200 years. This dialogue died in first beginnings and has never materialized. (61)[10]

For Scholem, in the wake of the Holocaust, the notion of a "supposedly indestructible intellectual community of the German with the Jewish ... was, on the level of historical reality, nothing more than a fiction, a fiction ... that was paid for too dearly."[11] (63) At the end of his letter, Scholem laments that it is only after the destruction of the German-Jewish community that Germans belatedly perceive the importance of the Jewish contribution to German culture and wish for a restitution of the German-Jewish dialogue. But, as Scholem implies, it is now too late to open the gates of the German Parnassus to Jewish participation. Scholem's characterization of a fictive cultural symbiosis between Germans and Jews (which the historian Dan Diner then amends in 1987 to "negative symbiosis") continues to influence representations of German-Jewish cultural history. Thus, at the beginning of the twenty-first century, when Willi Jasper constructs his comprehensive history of German-Jewish literature, which he titles, following Goldstein, "Deutscher-jüdischer Parnass" ["German-Jewish Parnassus"] he chooses to add a subtitle, "Literaturgeschichte eines Mythos" ["Literary History of a Myth"] to underscore the notional recognition of the Jewish contribution to literary history.

Almost a century after Moritz Goldstein's article and over forty years after Scholem's letter, Carl Djerassi adds another chapter to the complex cultural history of the German Parnassus and the tragic failure of the German-Jewish dialogue. With *Four Jews on Parnassus*, he expressly refigures the trope of Parnassus for his inner-Jewish dialogue and in the process overtly claims it as a site of German-Jewish intellectual contribution without either ignoring or effacing the

---

[10] "Ich bestreite, daß es ein solches deutsch-jüdisches Gespräch in irgendeinem echten Sinne als historisches Phänomen je gegeben hat. Zu einem Gespräch gehören zwei, die aufeinander hören, die bereit sind, den anderen in dem, was er ist und darstellt, wahrzunehmen und ihm zu erwidern. Nichts kann irreführender sein, als solchen Begriff auf die Auseinandersetzungen zwischen Deutschen und Juden in den letzten 200 Jahren anzuwenden. Dieses Gespräch erstarb in seinen ersten Anfängen und ist nie zustande gekommen." (7-8)

[11] "Die angeblich unzerstörbare geistige Gemeinschaft des deutschen Wesens mit dem jüdischen Wesen ... war, auf der Ebene historischer Realität, niemals etwas anderes als eine Fiktion, eine Fiktion, [...die] zu hoch bezahlt worden ist." (10)

brutal history of the Holocaust or feeling the need to justify the presence of his Jewish intellectuals on its exalted peaks. With his dialogue, Djerassi does not so much force open the gates of Parnassus, which, as we have seen, were resolutely closed to German Jews, as rethink and reconstitute Parnassus altogether, transforming not only the very premises of one's entry to the exalted site but also its composition and its *raison d'être*. Djerassi's imagined locale is no longer the mere resting place of genius, a static, calcified abode to be guarded against the profane hordes so that an exclusive notion of national literary poetic production might be preserved, but rather a sphere in which his intellectuals – Benjamin, Adorno, Scholem and Schönberg – engage actively and dialogically with each other and their intellectual past. His figures do not rest on their laurels in a poetic stupor; rather they query, provoke, examine, analyze, dissect, evaluate, criticize and interrogate. In short, they behave as intellectuals. On the one hand, they engage in what Djerassi's Benjamin calls "the rabbinical tradition of arguing and questioning and interpreting that is transmitted even after religion has disappeared from a formerly Jewish home" (144); at the same time, however, the content of their conversations is fully in keeping with their identity as German-Jewish and Austrian-Jewish secular intellectuals: death, love and sex, art, Jewish identity, intellectual and artistic legacy, and – not to be forgotten – their ambivalent love for German language and culture. As Benjamin says, "Kafka put it well when he talked about the three impossibilities of confronting us German Jewish writers: 'the impossibility of not writing; the impossibility of writing in German; and the impossibility of writing in another language'" (123). This impossibility of, desire for and necessity of German-Jewish existence, about which Kafka, another German-speaking Jew who must reside in another corner of Djerassi's Parnassus, expended so much anguish, constitute much of the dialogue on Parnassus; they are also the deep subject of Djerassi's book.

Like the figure of the Parnassus around which it centers, Djerassi's text is thus a mythical site of German-Jewish engagement. In his idiosyncratic integration of historical texts into an imagined dialogue, Djerassi creates a hybrid text that places the historical figures in conversation with each other in the idealized space of the Parnassus, a trope that, from its position on the other side of the divide that rends German-Jewish history, is refigured to represent the Jewish contribution to German culture in the twentieth century.

Works Cited

Diner, Dan. 1987. "Negative Symbiose: Deutsche und Juden nach Auschwitz." *Ist der Nationalsozialismus Geschichte? Zu Historisierung und Historikerstreit*. Ed. Dan Diner. Frankfurt am Main: Fischer.

Diner, Dan. 1990. "Negative Symbiosis: Germans and Jews After Auschwitz." *Reworking the Past: Hitler, the Holocaust and the Historian's Debate.* Ed. Peter Baldwin. Boston: Beacon. 251-261.

Djerassi, Carl. 2008. *Four Jews on Parnassus – A Conversation: Benjamin, Adorno, Scholem, Schönberg.* New York: Columbia UP.

Ehrendeutsch, Dichterecht [Adolf Wilhelm Schneider]. 1820. *Der Deutsche Parnass.* Meißen: Friedrich Wilhelm Goedsche.

Goethe, Wolfgang von. 1902. "The German Parnassus." *Poetical Works of J. W. von Goethe.* Vol. I. Ed. Nathan Haskell Dole. Boston: Francis A. Niccolls. 215-222.

Goethe, Johann Wolfgang von. 1977. "Sängerwürde." *Sämtliche Werke* Vol I. Munich: dtv. 712-719.

Goldstein, Moritz. 1912. "Deutsch-jüdischer Parnaß." *Der Kunstwart* 25: 281-294.

Jacoby, Daniel. 1893 "Goethes Gedicht: Deutscher Parnass." *Goethe-Jahrbuch* 14: 196-211.

Jasper, Willi. 2004. *Deutsch-jüdischer Parnass: Literaturgeschichte eines Mythos.* Berlin: Propyläen.

Minckwitz, Johannes. 1861. *Der neuhochdeutsche Parnaß, 1740 bis 1860: Eine Grundlage zum besseren Verständnisse unserer Litteraturgeschichte.* Leipzig: Arnoldische Buchhandlung.

Rist, Johann. 1652. *Neüer Teutscher Parnass.* Lüneburg: Johann und Heinrich Stern.

Scholem, Gershom. 1976. "Against the Myth of the German-Jewish Dialogue." *On Jews and Judaism in Crisis.* Ed. Werner J. Dannhauser. New York: Schocken. 61-64.

Sholem, Gershom. 1970. "Wider den Mythos vom deutsch-jüdischen Gespräch." *Judaica II.* Frankfurt am Main: Suhrkamp. 7-11.

Taylor, Bayard. 1895. *Studies in German Literature.* New York: G. P. Putnam's.

Wieland, Christoph Martin. 1790. "Der teutsche Parnaß." *Der neue teutsche Merkur* 1: 104-112.

The Idle Chatter of Ghosts:
On Carl Djerassi's *Four Jews on Parnassus*

Martin Jay

Staging an imaginary "dialogue of the dead" has been a temptation for a very long time indeed. In the *Odyssey*, Homer has Ulysses journey to the underworld to speak with Tiresias, Achilles and Agamemnon, among other illustrious shades, although Ajax, still smarting over losing Achilles' armor to Ulysses, refuses to join in the fun. In the 2nd century CE, the satirist Lucian, writing in Greek, composed his *Dialogues with the Dead* to go along with his dialogues with gods, seagods and courtesans. Combining the form of Plato's dialogues with the satirical content of the Cynic philosopher Menippus, he chastised human failure to realize the transience of wealth and fame. In subsequent centuries, Plutarch, Cicero, Augustine, Bocthius, and Erasmus developed the form of the colloquy to a high art. Not only were literary and philosophical dialogues invented, so too were scientific ones, a salient example being Galileo's *Dialogue Concerning the Two Chief World Systems* of 1632, which pitted the recently deceased Florentine Filippo Salviati, representing the new Copernican position, against a fictitious strawman named Simplicio, representing the old Aristotelian or Ptolemaic alternative. They debated their respective positions in the presence of an open-minded man of letters, the late Venetian Giovanni Francesco Sagredo.

Although falling into some disfavor over the years, the dialogue of the dead has not entirely disappeared, as shown by Peters Gay's *The Bridge of Criticism: Dialogues on the Enlightenment* of 1970, which appropriately has Lucian and Erasmus, themselves masters of the form, trading wit and wisdom with Voltaire. As recently as 2008, as if to show that the genre still has some life, a writer with the felicitous name of Baudelaire Jones updated Lucian by substituting more recent figures like Howard Hughes, John D. Rockefeller, Anna Nicole Smith, Clarence Darrow, Sigmund Freud, Michael Moore, Saddam Hussein, and Jack the Ripper for Lucian's cast of ancient characters. (Cf. Jones/Lucian)

Normally these dialogues have had an explicitly didactic purpose, presenting the conflicting ideas of their protagonists in such a way as to convince an audience of the rightness of one position or allowing them make up their own minds about the issues involved. But another motivation for adopting this perennially tempting genre is revealed by Carl Djerassi in his foreword to the lively dialogue he has composed to illuminate what he sees as an under-appreciated facet of the lives of four of the most powerful intellectuals of the past century: Walter Benjamin, Theodor Adorno, Gershom Scholem and Arnold Schönberg. "I have chosen the format of direct speech," he tells us, "to present an easily grasped and humanizing view" of these difficult and elusive figures (xv). Repeating the latter adjective at the end of the foreword, he explains again:

Since my purpose is to present a *humanizing* view of my four subjects rather than theoretical insight into their work, I feel that dialogic 'Intercourse of Caresses' [the term is a disparaging one from the Earl of Shaftesbury] rather than the more dispassionate third-person voice may be the most effective way of accomplishing this. (xviif.)

Having himself had to suppress the first person singular for many years in his own work as a scientist, Djerassi, who has had a second career writing novels, plays and autobiographies, tells us that he wants to avoid the conceit of omniscient narrator objectivity that characterizes most scholarly prose.

Humanization, we soon discover, means focusing on the personal lives of the four figures, with a special emphasis on their complicated and fraught relationships with the women in their lives: wives, former wives, lovers, and friends. To be fully human – indeed, perhaps human-all-too-human – is to be enmeshed in the emotionally laden, morally uncertain daily challenges that even the most glorious of heroes have to meet. It is to see the private figure behind the public mask. Although Djerassi does not explore the ways in which their intellectual or artistic achievements were endebted to these relationships, he nonetheless informs us that "all but one (Mathilde Schönberg) contributed to her husband's creative output." (20) But despite their importance, he notes, they have received scant notice in the voluminous literatures on them.

As a gesture of retrieval, drawing our attention to neglected figures whose gender undoubtedly contributed to their obscurity in the eyes of posterity (as well as the limited roles they were compelled to play in their own day), *Four Jews On Parnassus* also brings its male protagonists down from their elevated status on the mountain where the Greek muses dwelled and lets us see them in a new light. Perhaps it is worth remembering that the mid-19th-century French school of poetry called the Parnassians identified with figures like Charles Marie René Leconte de Lisle, had repudiated the looser forms of romantic poetry, with its excessive sentimentality and emotional intensity, in favor of more exact and controlled workmanship, even treating exotic and classical subjects with emotional detachment and formal rigor. By forcing his quartet of Jewish Parnassians to come down from their lofty perch through a dialogue with the women they left behind, Djerassi reminds us of the costs of focusing only on texts understood in formal terms and not on the lives out of which they emerged.

Humanization is a hard goal to dispute, especially when it compels us to ask about the underlying gender politics permitting the productivity of intellectual life. But in the case of the four protagonists chosen by Djerassi, there is a nagging question that needs to be addressed. For all four were themselves deeply skeptical of the very attempt to restore an allegedly human dimension, one that posited the primacy of interpersonal relations and emotional intensity, as the ground of intellectual work, at least as a possibility in their day. Each in his own way was deeply suspicious of contemporary attempts to humanize what had become a world that had made that goal, whatever it might mean, only an indirect and remote possibility. Indeed, in certain respects, they not only challenged attempts to realize it in the here and now as ideologically suspect, but also the very goal of humanization itself. Take, for example, Scholem's dismissive response to the attempt by Martin Buber to create a new existentialist theology of "I and Thou" in the Weimar era, a theology precisely of dialogue that emphasized the communicative, interpersonal closeness of God and man and man and man. From Scholem's point of view, it smacked of the vitalism that he so disliked in contemporary German philosophy, and showed that its author had never fully recovered from the *Erlebnismystik* of Buber's early years. Buber's attempt to draw on the legends of Hasidism, the everyday "Hasidic" life, he scorned in favor of returning to a rigorous reading of the theoretical texts of the Kaballah. "Buber's joy in life as it is and in the world as it is," he wrote in his 1961 essay "Martin Buber's Interpretation of Hasidism,"

seems to me a rather modern idea, and the Hasidic expressions seem to me to convey a totally different mood. They do not teach us to enjoy life as it is; rather they advise us – better: demand of us – to extract, I am tempted to say distill from 'life as it is' the perpetual life of God.... It is just this concept of abstraction in regard to joy and to the uplifting of the sparks to which Buber's interpretation of Hasidism objects. He does away with it because it runs counter to his essential interest in Hasidism as an anti-Platonic, existentialist teaching. (Scholem 1971, 240-241)

Classical Hasidic literature, Scholem insisted, "consistently treats the individual and concrete existence or phenomenon quite disdainfull." (Scholem 1971, 243) There was, to be sure, a moment in the history of Hasidism when the cult of the Zaddik, the saint figure of the movement, did emerge, but much to Scholem's disgust. In *Major Trends in Jewish Mysticism*, he noted with disdain what he saw as the "irrationalization of religious values":

Personality takes the place of *doctrine;* what is lost in rationality by this change is gained in efficacy. The opinions particular to the exalted individual are less important than his character, and mere learning, knowledge of the Torah, no longer occupies the most important place in the scale of religious values. (Scholem 1961, 344; Italics in the original)

In the case of Arnold Schönberg, any pressure to write music that was immediately accessible, music that appealed to the senses or feelings rather than the intellect, music that was emotionally moving, was rejected. Fully aware that his work was unpalatable to the common listener, he embraced the esoteric elitism of private performances rather than compromise his artistic integrity. Although he had gone through an earlier expressionist phase, both in his paintings and his music, when he moved to serialism in the 1920's, he deliberately abandoned the attempt to write music that was psychologically meaningful. Instead, he sought an implicit order beneath the surface, based on the principles of dodecophonic composition, that was inaccessible to the ear, a far cry from, say, the lush, evocative chromaticism of post-Wagnerian opera. Essence did not emphatically manifest itself in appearance, formal structure in phenomenal experience. When during his exile a Hollywood producer tried to flatter him by praising his "lovely music," Schönberg snapped back, "I don't write 'lovely' music." (cited in Ross 296) No populist, Schönberg reveled in the difficulty and unapologetic elitism of his compositions whose deep structure was never audible to a listener lacking in knowledge of the fundamental principles of his musical revolution. As Adorno was to put it, "Schönberg's music honors the listener by not making any concessions to him." (Adorno 1967, 154)

Likewise, for Walter Benjamin, the humanistic appeal to direct experience and interpersonal communication was also anathema. Exemplary in this regard is his attitude towards Buber's version of "I-Thou existentialism," which was even more hostile than Scholem's. He scornfully disdained Buber's praise of experiential intensity and immediacy as opposed to theoretical abstraction. As Scholem remembered it in the memoir he wrote of their friendship:

Benjamin was especially harsh in his rejection of the 'cult of experience,' which was glorified in Buber's writings of the time (particularly 1910 to 1917). He said derisively that if Buber had his way, first of all one would have to ask every Jew, 'Have you experienced Jewishness yet?' ... Benjamin said that Buber represented feminine thinking. In contrast to Gustav Landauer, who once said the same thing about Buber in an essay byway of praise, Benjamin meant it as a condemnation. (Scholem 1981, 29)

Benjamin's critique of the allegedly 'feminine' side of Buber's existentialist pathos of the dialogic self did not, however, lead to an embrace of a stereotypical masculine self: rational, boundaried, coherent, masterful and armored. Although he composed many texts that can be seen as drawing on his own life – *Berliner Kindheit um 1900, Einbahnstraße, Moskauer Tagebuch* – he was deeply skeptical of the ideal of a unified subject of experience. As Gerhard Richter has pointed out in his study of *Walter Benjamin and the Corpus of Autobiography*, "Benjamin rejects the notion of a continuous and self-identical subject that could account for its own multiplications and reconfigurations in the scene of writing." (Richter 36) He cites a letter Benjamin wrote to his friend Ernst Schoen in September 1919 in which he dismissed the idea of a human subject emerging from his texts, preferring instead to speak of a textual event "whose relation to a subject is as meaningless as the relation of any pragmatic-historical testimony (inscription) to its author." In short, language always goes beyond those who use it, and this is a good thing. As Adorno pointed out in his essay on Benjamin in *Prisms*: "From the start his thought protested against the false claim that man and the human mind are self-constitutive and that an absolute originates in them." The results could, Adorno conceded, be problematic, as

his target is not an allegedly over-inflated subjectivism but rather the notion of a subjective dimension itself. Between myth and reconciliation, the poles of his philosophy, the subject evaporates. Before his Medusan gaze, man turns into the stage on which an objective process unfolds. For this reason Benjamin's philosophy is no less a source of terror than a promise of happiness. (Adorno 1967, 235)

Adorno for his part was anxious not to let the evaporation of the subject happen without protest. But he, too, resisted the attempt to recover a robust and healthy version of subjectivity in the present age without revolutionary social transformation. "Today," he wrote in one of his essays on Schönberg, "subjectivity in its immediacy can no longer be regarded as the supreme category since its realization depends on society as a whole." (Adorno 1967, 157) Thus it was impossible to return to the days of, say, Beethoven, when subjective mastery of musical material was still possible. In philosophical and psychological terms, Adorno explicitly scorned the ideology of the coherent, active, bourgeois self. In *Minima Moralia*, just to take one example out of many, he wrote that

Hume, whose work bears witness in every sentence to his real humanism, yet who dismisses the self as a prejudice, expresses in this contradiction the nature of psychology as such. In this he even has truth on his side, for that which posits itself as 'I' is indeed mere prejudice, an ideological hypostatization of the abstract centers of domination, criticism of which demands the removal of the ideology of 'personality'. (Adorno 1974, 64)

Many other examples of the suspicion of straightforward humanism can be gleaned from the work of the main protagonists of *Four Jews on Parnassus,* but I hope the larger point has been made. Any attempt to humanize them by moving past their texts or compositions to the living people behind them, and doing so in particular by exploring their intimate relations with the women in their lives, comes up against the weight of their theoretically motivated protest against that very attempt. These were four dead intellectuals who were opposed in principle to dialogic immediacy. This is not, however, to say that it would be wiser to return to the third person, allegedly disinterested objectivism that Carl Djerassi has strained against in his own struggle to free himself from scientific prose and eschew any attempt at dialogic revivification. To this too they would have been deeply opposed.

Might it be better to adopt instead the cool detachment characteristic of the mood that dominated the middle years of the Weimar Republic in which they spent much of their formative years? I am referring here to the artistic movement that has sometimes been seen as emblematic of middle Weimar culture as a whole, the so-called *Neue Sachlichkeit* (new objectivity or new sobriety). Refusing the overheated emotionalism of the Expressionist years, it favored understated, functional realism, tinged with a cynical rejection of utopian hopes. Surfaces were more important than depths, objective truths more than subjective feelings and anonymity more than personal distinction. Instead of dialogic intimacy and interpersonal authenticity, it counseled wearing the mask of the dandy or the camouflage of the soldier.

In his recent book, *Cool Conduct: The Culture of Distance in Weimar Germany*, Helmut Lethen explores the personal style of those figures who rejected the cult of authenticity and sincerity characteristic of the Expressionist years, a style marked by deliberate character armoring, role-playing and artificiality, designed to render the threatened subject impervious to penetration from outside. Brecht's epic theater with its replacement of emotive expression by the stylized gesture is typical of the new mood, the epic theater that so attracted Benjamin to his playwright friend's aesthetic practice. Lethen argues that Benjamin's fascination with the allegorical culture of the baroque, a period like that of postwar Germany traumatized by war and the loss of metaphysical certainties, betokened his embrace of the spirit of the new objectivity. He compares his position with that of the anthropologist Helmut Plessner, whose endorsement of distance, abstraction, uprootedness and restlessness provides a phenomenology of the *Neue Sachlichkeit* sensibility.

Adorno was likewise very skeptical of the fetish of authenticity, which he identified with the dangerous cult of primitivist immediacy he saw in philosophical phenomenology and the music of composers like Stravinsky. As I have tried to show elsewhere, well before his postwar critique of the *Jargon of Authenticity*, he was denouncing the fraudulent search for origins, concreteness, roots and individual integrity, preferring to praise *mimesis*, interdependence and repetition instead. (see Jay) The alienation of the modern world cannot be undone, he insisted, by restoring contact with the allegedly raw 'life' beneath the distortions of social convention and economic abstraction. The antidote to the reign of the exchange principle is not the concrete authenticity of the unique individual.

But it is also true, and this is the larger point I want to make, a point which helps us to see the value of a dialogue of the dead of the sort written by Carl Djerassi, Adorno was also deeply skeptical of the *Neue Sachlichkeit* alternative to expressionist pathos and humanist self-worship. He accused it of surrendering to reification, the reification caused by capitalist modernization and the domination of the scientific worldview. Thus, when it came to characterizing Schönberg's move away from his atonal expressionism to the twelve-tone row, he resisted identifying it entirely with the *Neue Sachlichkeit*, however much it may have resembled it in certain respects.

For all his distrust of the humanist ideology of personal integrity, which he declared in its traditional, high liberal form "obsolete," he nonetheless could write in his postwar collection *Critical Models*,

the concept of personality cannot be saved. In the age of its liquidation, however, something in it should be preserved: the strength of the individual not to entrust himself to what blindly sweeps down upon him, likewise not to entrust himself to what blindly make himself resemble it. (Adorno 1998, 165)

What the alternative would look like Adorno could not say with confidence, but

at least something negative can be said about the concept of the real person. He would be neither a mere function of a whole, which is inflicted upon him so thoroughly that he cannot distinguish himself from it any more, nor would he simply retrench himself in his pure selfhood: precisely that is the form of a bad rootedness in nature that even now still lives on. (Adorno 1998, 165)

In his rejection of the cult of authentic selfhood rooted in an ideological notion of naturalism, Adorno opened up the question of relationality, which is inevitably the question of gender, the very question addressed in Carl Djerassi's dialogue. Perhaps more than any of the other figures in his quartet of Parnassians, Adorno understood the importance of gender relations as an indicator of the distortions of justice and the good life imposed by the present socio-economic order. In one of the most trenchant aphorisms of *Minima Moralia*, "Philomen and Baucis," Adorno pondered the costs of patriarchal marriage for both parties. Referring to the apparent catering of the wife to the husband's whims, he writes:

The patriarchal marriage takes its revenge on the master in the wife's indulgent considerateness, which in its ironic laments over masculine self-pity and inadequacy have become a formula. Beneath the lying ideology which sets up man as superior, there is a secret one, no less untrue, that makes him seem as inferior, the victim of manipulation, maneuvering, fraud. The hen-pecked husband is the shadow of him who has to go out to face the hostile world. (Adorno 1974, 173)

The pretensions of the husband to be the authoritarian patriarch are belied by his real status in that hostile world, which is that of exploited impotence. The dialectic of master and slave which Hegel outlined in *The Phenomenology* is played out in the household, in which both partners are lamed:

> As the repressed matriarch she becomes the master precisely where she has to serve, and the patriarch needs only to appear as such in order to be a caricature. This simultaneous dialectic of epochs has presented itself to individualistic eyes as the 'battle of the sexes.' Both opponents are in the wrong. In demystifying the husband, whose power rests on his money-earning trumped up as human worth, the wife too expresses the falsehood of marriage, in which she seeks her whole truth. (Adorno 1974, 173)

The implication of all this is that a mere reshuffling of gender relations in which equalization of roles and functions is the goal will not free either of them, for there is "no emancipation without that of society." (Adorno 1974, 173)

But, we have to wonder in conclusion, what are we to do until that state of full social emancipation is achieved? What kinds of gender relations can we abide or even foster in the era of what Adorno called in the subtitle of *Minima Moralia* "damaged life"? Adorno, to be sure, never thought there was an easy passage from theory to practice, especially in the current world, neither when it came to political activism nor to living an exemplary life. But it is nonetheless intriguing that his own personal situation as it has come down to us through the letters he wrote to his parents and other intimates suggests an unorthodox, even experimental approach to gender relations and family life. For example, in his correspondence with his mother, he was able to discuss his extramarital affairs with astounding candor. And for whatever reasons of her own, Gretel was also privy to these affairs, indeed seems to have countenanced them. Nor did he seem to mind when she had an affair as well. We can, of course, only speculate on the emotional cost of this kind of arrangement or perhaps fantasize about its benefits. But at least it suggests that the Adornos were unafraid of conventional mores and not beholden to the norms of traditional marriages. Insofar as Carl Djerassi's reconstruction of their relationship reminds of this fact, it is a very valuable exercise. Even if it may be a vain hope to restore personal immediacy and humanize the ideas by revealing the people who developed and defended them, it is enormously useful to be reminded that they were struggling with the contradictions of a society in which no perfect arrangement of gender relations were yet possible because of the larger forces at work in society. Even on the elevated heights of Parnassus, a dialogic intercourse of caresses comes up against the hard facts of life down on earth.

Works Cited

Adorno, Theodor W. 1967. *Prisms*. Trans. Samuel and Shierry Weber, London: Neville Spearman.

Adorno. Theodor W. 1974. *Minima Moralia: Reflections from Damaged Life*. Trans. E.P.H. Jephcott. London: NLB.

Adorno, Theodor W. 1998. *Critical Models: Interventions and Catchwords*, Trans. Henry W. Pickford. New York: Columbia UP.

Djerassi, Carl. 2008. *Four Jews on Parnassus – A Conversation: Benjamin, Adorno, Scholem, Schönberg*. New York: Columbia UP

Jay, Martin. 2006. "Taking on the Stigma of Inauthenticy: Adorno's Critique of Genuineness." *New German Critique*, 97 (Winter). 15-30.

Jones, Beaudelaire & Lucian of Samosata. 2008. *Dialogues of the Gods*. London: Black Box Press.

Lethen, Helmut. 2002. *Cool Conduct: The Culture of Distance in Weimar Germany*. Trans. Don Reneau. Berkeley: U of California.

Richter, Gerhard. 2000. *Walter Benjamin and the Corpus of Autobiography*. Detroit: Wayne State UP.

Ross, Alex. 2007. *The Rest is Noise: Listening to the Twentieth Century*. New York: Farrar, Strauss and Giroux.

Scholem, Gershom. 1971. "Martin Buber's Interpretation of Hasidism." *The Messianic Idea in Judaism and Other Essays on Jewish Spirituality*. Trans. Michael A. Mayer. New York: Schocken. 227-251.

Scholem, Gershom G. 1961. *Major Trends in Jewish Mysticism*. New York: Schocken.

Scholem, Gershom. *Walter Benjamin: The Story of a Friendship*. Trans. Harry Zohn. New York: Schocken.

Carl Djerassi:
A Literary Bibliography

This bibliography of Carl Djerassi's literary works is maintained at www.djerassi.com where it is continuously updated. Reproduction with kind permission of Carl Djerassi.
   *Asterisked entries denote books.

1983

1. *Paul Klee at Bard.* by Carl Djerassi (Poem) In *The Graphic Legacy of Paul Klee,* Edith C. Blum Art Institute, The Bard College Center, Annandale-on-Hudson, NY, pp. 10-11 (1983).

1984

2. *Why Are Chemists Not Poets?* by Carl Djerassi (Poem) In *Proceedings of the Alfred Benzon Symposium* 20, (P. Krogsgaard-Larsen, S. Broger Christensen, and Helmer Kofod, eds.) Munksgaard, Copenhagen, pp. 548-549 (1984).
3. *Morphine and Related Opioid Analgesics.* by Carl Djerassi (Poem) In *Proceedings of the Alfred Benzon Symposium 20,* (P. Krogsgaard-Larsen, S. Broger Christensen, and Helmer Kofod, eds.) Munksgaard, Copenhagen, pp. 448-449 (1984).
4. *Lingonberries Are Not Sufficient.* by Carl Djerassi (Poem) *Cumberland Poetry Review,* 3, 63 (1984).
5. *Concordance Browsing.* by Carl Djerassi (Poem) *Wallace Stevens Journal,* 8, 108 (Fall 1984).
6. *Pure Scientist, You Look With Nice Aplomb.* by Carl Djerassi Poem) *Wallace Stevens Journal,* 8, 109 (Fall 1984).
7. *Vocalissima.* by Carl Djerassi (Poem) *Wallace Stevens Journal,* 8, 110 (Fall 1984).
8. *Introduction to Paul Klee: Selections from the Djerassi Collection,* by Carl Djerassi, San Francisco Museum of Modern Art, San Francisco, California, pp. 2-5 (1984).

1985

9. *Two Egyptians Being Served by a Nazi.* by Carl Djerassi (Poem) *Wallace Stevens Journal,* 9, 52 (Spring 1985).
10. *The Twins.* by Carl Djerassi (Poem) *Wallace Stevens Journal,* 9, 118 (Fall 1985).

1986

11. Introduction to *Paul Klee: Figurative Graphics from the Djerassi Collection,* by Carl Djerassi, San Francisco Museum of Modern Art, San Francisco, California, p. 5 (1986).
12. *The Clock Runs Backwards.* by Carl Djerassi (Poem) *Kenyon Review,* 8, 91 (Winter 1986).
13. *Catalyst.* by Carl Djerassi (Poem) *Kenyon Review,* 8, 93 (Winter 1986).
14. *The Next Birthday.* by Carl Djerassi (Poem) *Kenyon Review,* 8, 94 (Winter 1986).
15. *Amour d'Arthropode (III).* by Carl Djerassi (Poem) *Current Contents,* 26, 4 (1986).
16. *Introduction* to *Quartet* (by Lewis Thomas). by Carl Djerassi, Arif Press, Berkeley, California, pp. 5-7 (1986).
17. *Cocksure.* by Carl Djerassi (Poem) *Metrosphere,* Issue 4, p. 84 (1986-87).
18. *Castor's Dilemma.* by Carl Djerassi (Short Story) *Hudson Review,* 39, 405 (1986).

1987

19. *My Island.* by Carl Djerassi (Poem) *The Midwest Quarterly,* 28, 214 (1987).
20. *Diary Entry (11 August 1983).* by Carl Djerassi (Poem) *Sands 1987,* 10 (1987).
21. *Wallace Stevens Was Right.* by Carl Djerassi (Poem) *Sands 1987,* 12 (1987).
22. *What's Tatiana Troyanos Doing in Spartacus' Tent?* by Carl Djerassi (Short Story) *Exquisite Corpse,* 5, 9 (1987).
23. *Maskenfreiheit.* by Carl Djerassi (Short Story) *The Crescent Review,* 5 (2) (1987).
24. Introduction to *Pattern and Process: Nature and Architecture in the Work of Paul Klee.* by Carl Djerassi San Francisco Museum of Modern Art, San Francisco, p. 5 (1987).
25. *Three Variations on a Theme by Callosobruchus.* by Carl Djerassi (Poem) *New Letters,* 54, 112 (Fall 1987).
26. *Godfather I.* by Carl Djerassi (Poem) *New Letters,* 54 (1), 113 (Fall 1987).

1988

27. *Why Are Chemists Seldom Poets?* by Carl Djerassi (Poem) *South Dakota Review,* 26, 128 (Summer 1988).
28. Introduction to *The In-Between-World of Paul Klee.* by Carl Djerassi, San Francisco Museum of Modern Art, San Francisco, p. 5 (1988).
29. *The Quest for Alfred E. Neuman* by Carl Djerassi (Memoir). *Grand Street,* 8, 167 (1988).
30. *The Toyota Cantos.* by Carl Djerassi (Short Story) *Frank* (Paris), No. 10, p. 90 (1988).
31.* *The Futurist and Other Stories.* by Carl Djerassi (Short Story Collection) Futura Publications (paperback), London (U.K.), 159 pp. (1988); Macdonald & Co. (hardback), London 1988…
32. *What's Tatiana Troyanos Doing in Spartacus' Tent?* by Carl Djerassi (Short Story), *Cosmopolitan* (U.K.), p. 233 (December 1988).
33. *My Island.* by Carl Djerassi (Poem) *Anthology of Magazine Verse,* 1986-1988 edition (Alan F. Pater, ed.), Monitor Book Company, Beverly Hills, California, pp. 136-137.

1989

34. *Cleansing My Doors of Perception.* by Carl Djerassi (Memoir) *Michigan Quarterly Review,* 28, 123 (Winter 1989).
35. "*Dear Mrs. Roosevelt.*" by Carl Djerassi (Memoir) *Hudson Review,* XLII (1), 61 (1989).
36. *My Very First Divorce.* by Carl Djerassi (Memoir) *Hudson Review,* XLII (1), 65 (1989).
37. *Castor's Dilemma.* by Carl Djerassi (German translation of short story) *Nachrichten aus Chemie,* 37, 403 (1989).
38. *Degas' Horse.* by Carl Djerassi (Memoir) *New Letters,* 55, (2), 13 (Winter 1988/89). (Winner of Dorothy Churchill Cappon Essay Award.)
39. *Spider at an Exhibition.* by Carl Djerassi (Poem) *Poet & Critic,* 20:3, p. 15 (1989).
40. *Kinshasha to Brussels* by Carl Djerassi (Memoir) *Exquisite Corpse,* 7:6-9, p. 22 (1989)

41. *Catalyst.* by Carl Djerassi (Poem) *The Kenyon Poet* (Anthology), p. 111 (1989).
42. *My Island.* by Carl Djerassi (Poem) *The Kenyon Poet* (Anthology), p. 112 (1989).
43. *The Next Birthday.* by Carl Djerassi (Poem) *The Kenyon Poet* (Anthology), p. 113 (1989).
44. *White House Enemy.* by Carl Djerassi (Memoir) *Negative Capability,* IX(1), p. 122 (1989).
45. *I Have Nothing Left to Say.* by Carl Djerassi (Poem) *Negative Capability,* IX, p. 65 (1989).
46.* *Cantor's Dilemma.* by Carl Djerassi, Doubleday, New York, 231 pp. (1989).

1990

47. *The Big Drop.* by Carl Djerassi (Memoir) *The Hudson Review,* XLII(4), p. 565 (1990).
48. *Freud and I.* by Carl Djerassi (Memoir) *The Southern Review,* 26 (1), p. 42 (1990).
49. *Wien, Wien, nur du allein* by Carl Djerassi (Memoir) (erroneously published as Wein, Wein, nur du allein...) *The Exquisite Corpse,* 8, (5-9), p. 37 (1990).
50.* *Cantor's Dilemma.* by Carl Djerassi (Novel) Macdonald, London (U.K.), 230 pp. (1990).
51.* *Steroids Made it Possible.* by Carl Djerassi (Scientific Autobiography) American Chemical Society Books, Washington, DC; 205 pages (1990).

1991

52. *Cash for Creativity.* By Carl Djerassi (Book Review:*The Need to Give* by Andrew Sinclair). *Times Lit. Suppl.,* January 4, 1991.
53. *A Scattering of Ashes.* by Carl Djerassi (Memoir) *The Hudson Review,* XLIII, No. 4, pp. 571-581 (1991).
54.* *Cantor's Dilemma.* by Carl Djerassi (Paperback Edition) Penguin Books, New York, 230 pages (1991).
55.* *Cantors Dilemma.* by Carl Djerassi Translated into German by Ursula-Maria Mössner). Haffmans Verlag, Zurich, 276 pages (1991).
56.* *The Clock Runs Backward.* by Carl Djerassi (Poetry Chapbook) Story Line Press, Brownsville, OR; 55 pages (1991).
57. *Some Forms of Art Patronage.* by Carl Djerassi *Bulletin of The American Academy of Arts and Sciences,* XLIV, (7), pp. 51-68 (1991).
58.* *Cantor's Dilemma.* by Carl Djerassi (Paperback Edition) Futura Publications, London, 230 pages (1991).
59. *Was macht Tatiana Troyanos in Spartakus Zelt?* by Carl Djerassi (Translated into German by Ingeborg Kuhn.) *Der Rabe,* Nr. 30, 419-423 (Haffmans Verlag, Zurich) (1991).
60.* *Der Futurist und andere Geschichten.* by Carl Djerassi (Translated into German by Ursula-Maria Mössner.) Haffmans Verlag, Zurich, 189 pages (1991).
61. *Der Psomophile.* by Carl Djerassi (German translation of short story.) *Spiegel Spezial,* 3, p. 98 (1991).
62. *How I Beat Coca-Cola: A Science Fiction.* by Carl Djerassi (Short Story). *South Dakota Review,* 29 (4), pp. 46-61 (Winter 1991).

1992

63. *Ein Rhinozeros hat meine Rolex geschluckt.* by Carl Djerassi (Memoir. Translated into German by Ursula-Maria Mössner.) *Der Rabe,* Nr. 30, pp. 26-30, Haffmans Verlag, Zurich (1992).
64. "*Was macht Tatiana Troyanos in Spartakus Zelt?*" (German translation of short story.) *Das Raben* Taschenbuch 160, Nr. 63, pp. 419-423, Haffmans Verlag, Zürich (1992).
65.* *The Pill, Pygmy Chimps, and Degas' Horse: An Autobiography.* by Carl Djerassi Basic Books, New York, 319 pp. (1992).
66.* *Le Dilemme de Cantor.* by Carl Djerassi (French translation by Josette Chicheportiche et Natalie Levisalles.) Editions Balland, Paris, France, 345 pp. (paperback edition) (1992).
67.* *Die Mutter der Pille – Autobiographie.* by Carl Djerassi (German translation by Ursula-Maria Mössner.) Haffmans Verlag, Zurich, 521 pp. (1992).
68. "*I Have Nothing Left to Say.*" by Carl Djerassi (Poem) *In: Life on the Line* (Anthology), ed. by Sue B. Walker & Rosaly D. Roffman. Negative Capability Press, Mobile, AL, p. 92 (1992).

1993

69 "*Managing Competing Interests: Chastity vs. Promiscuity*" by C. Djerassi in "*Ethics, Values, and the Promise of Science*" Forum Proceedings, Sigma Xi, the Scientific Research Society, Research Triangle Park, NC, 1993, pp. 31-45.
69a. "*Die Uhr läuft rückwärts*" (The Clock Runs Backward [Poem]) In: *Der Rabe,* Nr. 35, pp. 94-96, Haffmans Verlag, Zurich (1993).
70. "*Die Uhr laüft rückwärts*" (*The Clock Runs Backward* by Carl Djerassi [Poem]) in *Kiesstrasse Zwanzig Uhr* (ed. Jürgen Lentes), Frankfurt, pp. 180-185 (1993).
71.* *Das Bourbaki Gambit.* by Carl Djerassi German translation by Ursula-Maria Mössner. Haffmans Verlag, Zurich, 278 pp. (1993).
72.* *The Pill, Pygmy Chimps, and Degas' Horse: An Autobiography.* by Carl Djerassi Basic Books, New York, paperback edition, 319 pp. (1993).
73.* *Cantors Dilemma* by Carl Djerassi (German translation by Ursula-Maria Mössner). Wilhelm Heyne Verlag, Munich, paperback edition, 287 pp. (1993).
74.* *Il futurista e altri racconti* by Carl Djerassi (Italian translation by Adriana Crespi Bortolini). Sellerio editore Palermo, 174 pp. (1993).
75.* *Der Futurist und andere Geschichten.* by Carl Djerassi (German translation by Ursula-Maria Mössner). Wilhelm Heyne Verlag, Munich, paperback edition, 188 pp. (1993).

1994

76.* *Cantor's Dilemma* by Carl Djerassi (Japanese translation). Bungeishunju, Tokyo, 379 pp. (1994).
77.* *El Dilema de Cantor* by Carl Djerassi (Spanish translation by Maria Urquidi). Fondo de Cultura Economica, Mexico, 194 pp. (1994).
78. "*Wachstum und die Folgen*" by Carl Djerassi in *Kleine Weltbevölkerungs-Fibel,* p. 31, 1994. Haffmans Verlag, Zürich.

79.* *La Pillola, gli Scimpanzé Pigmei el Cavalli di Degas* by Carl Djerassi (Italian translation by Marco Ferrari). Garzanti Editore s.p.a., Milano, 389 pp., (1994).
80.* *The Bourbaki Gambit.* by Carl Djerassi. University of Georgia Press, Athens, Georgia, 230 pp. (1994).
81.* *Marx, verschieden* by Carl Djerassi. (German translation of *Marx, deceased* by Ursula-Maria Mössner). Haffmans Verlag, Zurich. 276 pages (1994).
82.* *From the Lab into the World: A Pill for People, Pets, and Bugs* (Collected essays) by Carl Djerassi, Am. Chem. Soc. Books, Washington, D.C., 230 pp. (1994).
83. "*1991 Kein Jahr der Pille*" by Carl Djerassi in *Der Rabe*, No. 42, pp. 169-184. Haffmans Verlag, Zurich (1994).

1995

84. "*The Physics of a Man's Emotional Life*" by Carl Djerassi (book review of *Good Benito* by Alan Lightman) *San Francisco Chronicle* 1.29.95
85.* *Djerassi: De La Chimie des Hormones a La Pilule.* (French translation of autobiography), Editions Belin, Paris, 415 pp. (1995).
86. "*Die Suche nach Alfred E. Neuman*" by Carl Djerassi in *Der Rabe* No. 43, pp. 137-144. Haffmans Verlag, Zurich (1995).
87. "*Noblesse Oblige*" by Carl Djerassi in *Freundin* (Munich semimonthly), No. 24 (Nov. 8, 1995) pp. 171-180.
88.* *Das Bourbaki Gambit* by Carl Djerassi (German translation by Ursula-Maria Mössner) Wilhelm Heyne Verlag, Munich, paperback edition, 283 pp. (1995).

1996

89.† *Die Mutter der Pille* by Carl Djerassi (German translation by Ursula-Maria Mössner) Wilhelm Heyne Verlag, Munich, paperback edition, 539 pp. (1996).
90.* *La Píldora, Los Chimpancés Pigmeos y el Caballo de Degas* by Carl Djerassi (Spanish translation by Roberto Elier and Hugo Martínez Moctezuma) Fondo de Cultura Economica, Mexico, paperback edition, 381 pp. (1996).
91. "*Darf ich Sie Kosten?*" ("*May I Taste You?*") by Carl Djerassi in *Die Neue Klassische Sau (Handbook of Literary High-Eroticism)*, pp. 298-300, Haffmans Verlag, Zurich, 1996.
92.* *Marx, Deceased*, by Carl Djerassi The University of Georgia Press, Athens, 218 pp., (1996).
93.* *Bourbaki et spil om magt* by Carl Djerassi (Danish translation by Ingeborg Christensen), Fremad, Copenhagen, 264 pp., (1996).
94.* *The Bourbaki Gambit*, by Carl Djerassi (paperback edition), Penguin-USA, New York, 230 pp. (1996).
95.* *Menachems Same* by Carl Djerassi (German translation by Ursula-Maria Mössner), Haffmans Verlag, Zurich, 245 pp. (1996).
96.* *Cantor's Dilemma* by Carl Djerassi (Chinese translation by Jenny Yang et al and revised by Chai-li Wu), Unitas Publication Company, Taipei, 283 pp. (1996)
97.* *El Gambito de Bourbaki* by Carl Djerassi (Spanish translation by Leticia Garcia Urriza) Fondo de Cultura Economica, Mexico, paperback edition, 240 pp. (1996).

1997

98. *From the Pill to the Pen*, by Carl Djerassi *Contemporary Authors Autobiography Series*, Volume 26, pp. 85-107. Gale Research, Detroit (1997).
99.* *Menachem's Seed*, by Carl Djerassi, University of Georgia Press, Athens, GA., 196 pp. (1997).
100. *ICSI*, by Carl Djerassi (Sample scene from Play) "Reduce to the Max." – Micro Compact Car AG, Biel, Switzerland (1997) in English, German, French, Spanish, Italian and Dutch promotion for Mercedes Smart Car.
101. *Passionate Minds*, Chapter 2 titled *Blemished Heroes*, pp. 8-16, (by Lewis Wolpert and Alison Richards), Oxford University Press, Oxford (1997).

1998

*102. *Von der Pille zum PC*, by Carl Djerassi (German translation by Ursula-Maria Mössner), Haffmans Verlag, Zurich, 80 pp. (1998).
*103. *NO*, by Carl Djerassi (German translation by Ursula-Maria Mössner), Haffmans Verlag, Zurich, 335 pp. (1998).
104. In Retrospect: *The Struggles of Albert Woods*, by Carl Djerassi, Nature, 392, 244 (1998).
105. *Ethical Discourse by Science-in Fiction*, by Carl Djerassi Nature, 393, 511-513 (1998).
*106. *Dilemata na Kantor* by Carl Djerassi (Bulgarian translation by R. Radeva) Marin Drinov Publ. House, Bulg. Academy of Sciences, Sofia, 185 pp. (1998).
107. *Was macht Tatiana Troyanos in Spartakus' Zelt?* by Carl Djerassi in *Das Grosse Heyne Jubiläums Buch*, Heyne Verlag, Munich, pp. 565-568 (1998).
*108. *Menachem's Seed* by Carl Djerassi (paperback), Penguin-USA, New York, 196 pp. (1998).
*109. *NO,* by Carl Djerassi University of Georgia Press, Athens, GA, 276 pp. (1998).
110. *The Pill is the Pill is the Pill*, by Carl Djerassi *Los Angeles Times* Commentary, September 10, 1998, p. B9.
111. *Surely you're joking* by Carl Djerassi (Book review of *Dancing Naked in the Mind Field* by Kary Mullis), *New Scientist*, September 21, 1998, p. 51.
112. *Dylemat Profesora Cantora* by Carl Djerassi (Polish translation of short story by Janusz Stanny), Magazyn Gazety NR 41 (292) 1998, pp. 74-79.
113. *Science-in-fiction ist nicht Science Fiction: Ist sie Autobiographie?* by Carl Djerassi in "Fiction in Science – Science in Fiction. Zum Gespräch zwischen Literatur und Wissenschaft" (ed. W. Schmidt-Dengler), Verlag Hölder-Pichler-Tempsky, Vienna, pp. 71-104, 1998.

1999

114. *On the Pill: A Social History of Oral Contraceptives, 1950-1970* by Carl Djerassi (Review of book by Elizabeth Siegel Watkins), *New England Journal of Medicine*, February 11, 1999, Volume 340, pp. 485-486.
115. *Der entmachtete Mann oder: Sex im Zeitalter der technischen Fortpflanzung*, by Carl Djerassi in DIE ZEIT (Hamburg), No. 27, July 1, 1999, p. 28

*116. *O Dilema de Cantor* by Carl Djerassi (paperback), (Portuguese translation of *Cantor's Dilemma* by Flávia Terra Cunha) Editora Nova Fronteira (Rio de Janeiro, Brazil), 292 pp. (1999).
117. *Under the covers or under a lens* by Carl Djerassi (book review of Designing Babies by Roger Gosden), Nature, 22 July 1999, No. 6742, p. 327.
*118. *The Bourbaki Gambit,* by Carl Djerassi (Chinese translation by Jih-Li Chang and Chai-Li Wu), Unitas Publication Company, Taipei, 321 pp. (1999)
*119. *Menachem's Seed,* by Carl Djerassi (Chinese translation by Ding-Chi Chang), Unitas Publication Company, Taipei, 287 pp. (1999)
*120. *NO,* by Carl Djerassi (Chinese translation by Zy-Yin Chiu), Unitas Publication Company, Taipei, 343 pp. (1999)
121. "*The Century of A.R.T.*" by Carl Djerassi in "*Predictions*" (S. Griffiths, ed.), Oxford University Press, Oxford, 1999, pp. 76-84.
122. "*Useful Work*" by Carl Djerassi in "*Profession 1999*" (P. Franklin, ed.), The Modern Language Association of America, 1999, pp. 36-48.

2000

*123. *Unbefleckt*, by Carl Djerassi (German translation by Bettina Arlt), Haffmans Verlag, Zurich, 113 pp. (2000).
124. "*Erektion*" by Carl Djerassi in "*Die Allerneueste klassische Sau,*" Haffmans Verlag, Zurich, 1999, pp. 75–84.
125. "*Erregung in der Oper*" by Carl Djerassi in "*Die Allerneueste klassische Sau,*" Haffmans Verlag, Zurich, 1999, pp. 290-293
*126. *An Immaculate Misconception* by Carl Djerassi (paperback), Imperial College Press, London, 134 pp. (2000).
*127. "*Obefläckad*" by Carl Djerassi (translated into Swedish by Lolo Amble), Bookhouse Publishing, Stockholm 2000, 74 pp.
128. "Sex and Fertilization: Ready for Divorce?" by Carl Djerassi in NEXT SEX -Ars Electronica 2000 (ed. by G. Stocker & Christine Schöpf), Springer Verlag, Vienna & New York, 2000, pp. 45-52 (English), pp. 53-61 (German).
*129. *NO.* by Carl Djerassi (Novel – Paperback Edition) Penguin Books, New York, 2000, p. 276.
*130. *Wie ich Coca-Cola schlug,* by Carl Djerassi (German translation by Ursula-Maria Mössner), Haffmans Verlag, Zurich, 216 pp. (2000).
131. "Science-in-Theatre: From *Oxygen* (theater)"by Carl Djerassi and Roald Hoffmann in *Connecting Creations/Science-Technology-Literature-Arts* (ed. by Margery Arent Safir), Centro Galego de Arte Contemporánea, Santiago de Compostela, 2000, pp. 321-332 (English).
132. "La ciencia en el teatro: *Oxígeno*" por Carl Djerassi y Roald Hoffmann en *Conectando creaciones/Ciencia-Tecnología-Literatura-Arte* (Editado por Margery Arent Safir), Centro Galego de Arte Contempor Contemporánea, Santiago de Compostela, 2000, pp. 351-363 (Spanish).
*133. *Marx, el difunto.* by Carl Djerassi (Paperback Edition) (Spanish translation by Leticia García Urriza), Fondo de Cultura Económica, Mexico 2000, 228 pp.

*134. *La píldora anticonceptiva 40 AÑOS de impacto social*, Publicaciones M.V., Bogota Colombia, 2000, Foreword by Carl Djerassi, pp. 6-9. (Spanish)

*135. *La pílula anticoncepcional 40 ANOS de impacto social*, Publicacones M.V., Bogota Colombia, 2000, Foreword by Carl Djerassi, pp. 6-9. (Portuguese)

2001

*136. *Oxygen* by Carl Djerassi and Roald Hoffmann, (paperback), Wiley-VCH, Weinheim, 2001, 119 pp.

137. From *Oxygen A Play in Two Acts* by Carl Djerassi and Roald Hoffmann, The Kenyon Review & Stand Magazine, Volume XXIII Number 2, Spring 2001, Stand, Volume 2 (4)/3(1), March 2001 (paperback), pp. 221-238.

*138. *La Semilla de Menachem*, by Carl Djerassi (paperback edition) (Spanish translation by Marcela Pimentel), Fondo de Cultura Economica, Mexico, 2001, 217 pp.

*139. *This Man's Pill-Sex, Die Kunst und Unsterblichkeit*, (hardcover) by Carl Djerassi (German translation by Ursula-Maria Mössner) Haymon Verlag, Innsbruck, 2001, 235 pp.

*140. *This Man's Pill – Reflections on the 50th Birthday of the Pill*, by Carl Djerassi (hardcover) Oxford University Press, London, New York, 2001, 308 pp.

*141. *Oxygen* by Carl Djerassi and Roald Hoffmann, (paperback) (German translation by Edwin Ortmann), Wiley-VCH, Weinheim, 2001, 137 pp.

*142. *Die Mutter der Pille* (paperback) by Carl Djerassi (German translation by Ursula Maria Mössner) Diana Verlag, Munich, Zurich, 2001, 539 pp.

*143. *Cantor's Dilemma*, by Carl Djerassi (Japanese translation by Michio Nakamori), Kodansha, Tokyo, 2001, 317 pp.

*144. *The Bourbaki Gambit*, by Carl Djerassi (Japanese translation by Ryoji Nakanishi), Kodansha, Tokyo, 2001, 366pp.

*145. *Marx, Deceased,* by Carl Djerassi (Japanese translation by Ryoji Nakanishi), Kodansha, Tokyo, 2001, 366pp.

*146. *NO,* by Carl Djerassi (Japanese translation by Naemi McPherson), Kodansha, Tokyo, 2001, 444pp.

*147. "La Pildora de este hombre: Reflexiones en torno al 50 aniversario de la Pildora", by Carl Djerassi (translation by Abdiel Macias), Fondo de Cultura Economica, Mexico 2001, 235 pp.

148. "OXYGEN – A SYNOPSIS" by Carl Djerassi and R. Hoffman in *Oxygen: Wissenschaft im Theater* (ed. by M-D. Weitze and D. Champion), Deutsches Museum, Munich, 2001, pp. 36-39.

149. "WIE ENTSTAND OXYGEN?" by Carl Djerassi in *Oxygen: Wissenschaft im Theater* (ed. by M-D. Weitze and D. Champion), Deutsches Museum, Munich, 2001, pp. 52-61.

150. "*Sex und Unsterblichkeit*" ("Sex and Immortality") by Carl Djerassi in *Sex: Vom Wissen und Wünschen*, Deutsches Hygiene Museum, Hatje Cantz Verlag, Ostfildern-Ruit 2001, pp. 197-217.

151. "*Wien, Wien, Nur Du Allein*" by Carl Djerassi in *Das kann einem nur in Wien passieren* (Ruth Wodak, Editor), Czernin Verlag, Vienna 2001, pp. 41-43.

2002

*152. *Stammesgeheimnisse,* (paperback containing *Cantors Dilemma* & *Das Bourbaki Gambit*) by Carl Djerassi (German translation by Ursula-Maria Mössner), Haymon-Verlag, Innsbruck, 2002, 415 pp.

*153. *Marx, Deceased* by Carl Djerassi (paperback) (Chinese translation), Unitas Publication Company, Taipei, 308 pp. (2002).

*154. Paul Klee: Masterpieces of the Djerassi Collection, edited by Carl Aigner and Carl Djerassi. Prestel Verlag, Munich, 2002, 151 pages. Also published in German.

155. "*Diary Entry (11 August 1983)*" poem reprinted in "Foreword" by Carl Djerassi in *From Classical to Modern Chemistry* (P.J.T. Morris, editor), RSC/Science Museum, London, 2002, pp. v-vii.

156. "*Staccioli at Djerassi*" by Carl Djerassi in "*Mauro Staccioli in California*" (Francesca Pola, editor), Instituto Italiano di Cultura, Los Angeles, 2002, pp. 19-21.

*157. *ICSI-Sex in Zeitalter der technischen Reproduzierbarkeit/ICSI-Sex in the Age of Mechanical Reproduction* by Carl Djerassi (paperback contains both German and English versions) (German translation by Bettina Arlt), Deutscher Theaterverlag, Weinheim, 2002, 99 pp. plus compact disc.

158. "*Metastasen*" by Carl Djerassi in "Bürokratie und Subversion" (ed. W. Grünzweig, M. Kleiner & W. Weber), pp. 5-8, LIT Verlag, Münster, 2002.

159. "*Oxygen*" by Carl Djerassi and Roald Hoffmann (Korean translation by Duckhwan Lee), Freedom Academy, (Korean copyright by Korean Chemical Society), Seoul 2002, 232 pp.

*160. *Inmaculada concepción furtiva: El sexo en la era de la reproducción mecánica*, by Carl Djerassi (paperback edition) (Spanish translation by Marcela Pimentel), Fondo de Cultura Economica, Mexico, 2002, 120 pp.

161. *Contemporary Science-in-Theatre: A rare Genre* by Carl Djerassi, Interdisciplinary Science Reviews, 27 (3), 193 (2002).

2003

162. "*What's Popular in Science?*" (Closing Lecture) by Carl Djerassi (German language), Nova Acta Leopoldina NF 87, Nr. 325, pp. 243-246 (2003).

*163. "*This Man's Pill – Reflections on the 50th Birthday of the Pill,*" by Carl Djerassi (paperback) Oxford University Press, London, New York, 2003, 308 pp.

*164. "*ICSI—Sex in the Age of Mechanical Reproduction*" (Chinese translation by Hwang Kwei), Fembooks, Taipei, 2003, 105pp. plus compact disc.

165. "*From Oxygen*" by Carl Djerassi and Roald Hoffmann in *Writing on Air*, (David Rothenberg and Wandee J. Pryor, Editors), Massachusetts Institute of Technology, Cambridge, Massachusetts, pp. 105-113 (2003).

*166. "*Kalkül/Unbefleckt: Zwei Theaterstücke aus der Welt der Wissenschaft,*" by Carl Djerassi, Haymon Verlag, Innsbruck 2003, pp. 137.

*167. "*Newton's Darkness: Two Dramatic Views*" (by C. Djerassi and D. Pinner), Imperial College Press, London, 2003, pp. 184.

*168. "*NO – a pedagogic wordplay for 3 voices*" by Carl Djerassi and P. Laszlo (paperback contains German, French and English versions; German translation by Bettina Arlt), Deutscher Theaterverlag, Weinheim, 2003, 160 pp. plus compact disc.
*169. *Ossigeno* (by Carl Djerassi and R. Hoffmann) (paperback)(Italian translation of Oxygen by Daniela Majerna), Clueb, Bologna 2003, pp. 108.
*170. *Il Dilemma Di Cantor* by Carl Djerassi (paperback) (Italian translation of Cantor's Dilemma by Federico Maria Rubino), Di Renzo, Rome, 2003, pp. 340.
*171. *Oxígeno* by Carl Djerassi and Roald Hoffmann (paperback) (Spanish translation by Abdiel Macías), Fondo de Cultura Economica, Mexico City, 2003, pp. 100.
*172. *NO* by Carl Djerassi (paperback) (Spanish translation by Ana Pelida Rull) Fondo de Cultura Economica, Mexico City, 2003, pp. 306.
*173. "*How I Beat Coca-Cola and Other Stories*" (paperback) (Taiwanese translation by Huai-tung Peng) Unitas, Taipei, 2003, pp. 196.
174. "*Waiting for the Man*" by Carl Djerassi in *The Times Literary Supplement* (United Kingdom) of October 31, 2003, p. 25, book review of *The Male Pill* by Nelly Oudshoorn.
175. "*Thoughts on Turning Eighty*" (*Nur Wissenschafter verleugnen öffentlich ihr Streben nach Ruhm*) by Carl Djerassi in *Der Standard*, Vienna, October 29, 2003, p. 25.
176. "*Kann man Forschung verbieten?*" by Carl Djerassi in *Conturen 01/03*, Austria Perspektiv, Vienna, 2003, pp. 52-73.
*177. "*Oxygen*" by Carl Djerassi and Roald Hoffmann (paperback) (Chinese translation by Zhong Aimin), Shanghai Scientific & Technological Education Publishing House, 2003, pp. 150.
*178. "*Oxygène*" by Carl Djerassi and Roald Hoffmann (paperback) (French translation by Aimée & Jean-Michel Kornprobst, Presses Universitaires du Mirail (PUM), Toulouse, 2003, pp. 151.

2004

179. "*Science-in-Fiction: Literary Contraband?*" by Carl Djerassi in *Science, Technology, and the Humanities in Recent American Fiction*, (Peter Freese and Charles B. Harris, editors) (paperback) Die Blaue Eule, Essen, 2004, pp. 181-199.
180. "'*Science-in-Fiction' and 'Science-in-Theatre' as Pedagogic Tools, An Anglo-German Presentation*" by Carl Djerassi in *Facetten einer Wissenschaft* (A. Müller, H.-J. Quadbeck-Seeger, E. Diemann, Editors), Wiley-VCH, Weinheim, 2004, pp. 299-311.
*181. "*Ego: Roman und Theaterstück*" by Carl Djerassi (paperback)(German translation by Ursula-Maria Mössner) Haymon Verlag, Innsbruck, 2004, pp. 285.
182. "*Science-in-Fiction: Literary Contraband?*" by Carl Djerassi in *The Holodeck in the Garden: Science and Technology in Contemporary American Fiction,* (Peter Freese & Charles B. Harris, editors) Dalkey Archive Press, Illinois State University, 2004, pp. 322-333.
183. "*Smuggling Science onto the Stage or Page*" by Carl Djerassi in *Modern Biology & Visions of Humanity* (paperback), Multi-Science Publishing Company Ltd, Essex UK, 2004, pp. 201-214.

184. *"Introdurre la scienza nella pagina o sul palcoscenico"* by Carl Djerassi in *Biologia Moderna e Visioni Dell'Umanità* (Italian translation) Casa Editrice Università La Sapienza, Roma, 2004, pp. 207-221.
185. *"Le traffic de science dans les pages ou sur la scène"* by Carl Djerassi in *Biologie moderne & Visions de l'Humanité* (French translation) De Boeck & Larcier, Brussels, 2004, pp. 215-230.
*186. *"Oxigênio"* by Carl Djerassi and Roald Hoffmann (Portuguese-Brazilian translation by Juergen Heinrich Maar), Vieira & Lent casa editorial, Rio de Janeiro, 2004, pp. 144.
187. *"Literarisches Schmuggeln"* by Carl Djerassi in *Laborjournal*, No. 7-8 (special 10th anniversary issue), pp. 52-52 (2004).
*188. *Dalla pillola alla penna* by Carl Djerassi (Italian translation by Maria Pia Felici), Di Renzo, Rome, 2004, pp. 109.
*189. *ICSI: Il sesso nell'epoca della riproduzione meccanica* by Carl Djerassi (Italian translation by Maria Pia Felici together with English original text), Di Renzo, Rome, 2004, pp. 85.
*190. *"Sex, Art, and Immortality: This Man's Pill: Reflections on the 50th Birthday of the Pill"* by Carl Djerassi, (Chinese translation), Unitas Publication Company, Taipei,. (2004), 273 pp.
191. *"Competing visions: America in 2023"* by C. Djerassi in *The United States in Global Contexts: American Studies after 9/11 and Iraq* (Walter Grünzweig, ed.), Lit Verlag, Münster 2004, p. 33.
*192. *"Cantor's Dilemma"* by Carl Djerassi, (Chinese translation by Huang Qun), Baihua Literature and Art Publishing House, Shanghai. 2004, pp. 294.

2005

*193. *"Operazione Bourbaki"* by Carl Djerassi (Italian translation by Francesca Garafoli), Di Renzo, Rome, 2005, pp. 234.
194. *"Phallacy"* by Carl Djerassi (excerpts from play, translated by Ursula-Maria Mössner), in *Nachrichten aus der Chemie,* 53, 407-410 (2005).
195. *"Verwirrte Moralisten"* by Carl Djerassi in *Frankfurter Allgemeine Zeitung*, Feuilleton section, p. 35, June 18, 2005.
*196. *"The Pill, Pygmy Chimp's, and Degas' Horse"* by Carl Djerassi, (Chinese translation by Yao Ning), in "Philosopher's Stone" series, Shanghai Scientific & Technological Education Publishing House, Shanghai, 2005, 401 pp.
*197. *"A Tabletta: a fogamzásgátlás atyjának önéletrajza"* by Carl Djerassi. Hungarian translation (by Ágnes Borbély) of *"The Pill, Pygmy Chimps, and Degas' Horse"*, EDGE-2000, Budapest, 2005, pp. 397.
*198. *"Oxigênio"* by Carl Djerassi and Roald Hoffmann (Portuguese translation by Manuel João Monte), Universidade do Porto, Porto, 2005, pp. 175.
*199. *"Aufgedeckte Geheimnisse",* (containing *Menachems Same & NO – Zwei Romane aus der Welt der Wissenschaft*) by Carl Djerassi (German translation by Ursula-Maria Mössner), Haymon-Verlag, Innsbruck, 2005, 443 pp.

2006

200. *"Naturwissenschaften in das Bewusstsein schmugglen"* by Carl Djerassi in *"Von Science zu Fiction: Wissenschaft mit anderen Worten"* (ed. by E. Krottenthaler und C. v. See), S. Hirzel Verlag, Stuttgart 2006, pp. 16 – 19.

*201. *"Calcolo"* by Carl Djerassi (Italian translation of *Calculus* by Leonarda Anselmo), Di Renzo, Rome, 2006, pp. 93.

*202. *"Phallstricke/ Tabus: Zwei Theaterstücke aus den Welten der Naturwissenschaft und der Kunst"* by Carl Djerassi (translated by Ursula-Maria Mössner), Haymon Verlag, Innsbruck 2006, pp. 197.

*203. *"NO"* by Carl Djerassi (Italian translation of *NO* by Maria Pia Felici), Di Renzo, Rome, 2006, pp. 287.

204. *"La Comédie de l'Oxygène"* by Carl Djerassi and Roald Hoffmann in *Alliage 57-58 Science & Littérature*, Nice 2006, pp 107-113.

205. *"Aus schwierigen Verhältnissen"* by Carl Djerassi in ZEITWISSEN (Hamburg), No. 3, 2006, pp. 84-89.

206. *"Im Labor wird aus dem Zellklumpen sowieso kein Baby"* by C. Djerassi in *Frankfurter Allgemeine Zeitung*, Feuilleton section, p. 31, August 4, 2006.

207. *"A concrete example: When the cracks begin to show"* by J.-P. Boon, J. Casti, C. Djerassi, J. Johnson, A. Lovett, T. Norretranders, V. Patera, C. Sommerer, R. Taylor & S. Turner, NATURE, 444 (Nov. 2), 122 (2006).

208. *"Lob der Polygamie"* (Lichtenberg Medal Address) by C. Djerassi in Jahrbuch der Akademie der Wissenschaften zu Göttingen 2005 (publ. 2006), pp. 20–25 *(Festvortrag bei der öffentlichen Sitzung am 18. Nov. 2005, Lichtenberg Medaille).*

209. *"Play Scene: Are sex and fertilisation ready for divorce?"* by C. Djerassi, THE LANCET, 368, *Medicine and Creativity special Issue*, Dec. 2006, pp. S55-S57.

2007

210. *"When is 'Science on Stage' really Science?"* by C. Djerassi, AMERICAN THEATRE, Vol. 24, No. 1 (January) pp. 96-103 (2007).

211. *"Medical Milestones The pill: Emblem of liberation,"* by C. Djerassi, BRITISH MEDICAL JOURNAL, 334 (suppl.) s15, 2007.

212. *"When acting speaks louder than words: Science on Stage"*, book review by C. Djerassi, PHYSICS TODAY, 60 (No. 2), February 2007, pp. 63-64.

*213. *"NO"* by C. Djerassi (Chinese translation of *NO* by Li Hui), Shanghai People's Publishing House, Shanghai, 2007, pp. 300.

214. *Can research be forbidden"* By C. Djerassi in PROCEEDINGS OF THE WORLD CONGRESS FOR FREEDOM OF SCIENTIFIC RESEARCH (Rome, Feb. 16-18, 2006), Editore Cooper srl, Rome, 2007, pp. 69-83.

215. *"Olympiade der Chemiker"* by C. Djerassi in FRANKFURTER ALLGEMEINE SONNTAGSZEITUNG, 12 August 2007, Nr. 32, p. 26.

216. *"Leusuceutics: Opportunity or Curse?"* by C. Djerassi, Chem/Med/Chem, 2, 533 (2007)

217. *"Athletes and Steroids: A Chemist's Forecast"* In SAN FRANCISCO CHRONICLE, 7 October 2007, Section E, p. 1.

218. *"Paul Klee: La Visión de un Coleccionista"* by C. Djerassi in PAUL KLEE: LA INFANCIA EN LA EDAD ADULTA, Centro Atlantico de Arte Moderna, Las Palmas de Gran Canaria, 2007, pp. 70-89 (in English & Spanish).

2008

219. *"Science plays on stage"* A Reply by C. Djerassi.PHYSICS TODAY, 61 (No. 2), February 2008, pp. 11-12.
220. *"Sport Doping—a double blind proposal"* by C. Djerassi, Chem/Med/Chem, 3, 361 – 362 (2008)
221. *"Paul Klee: Aus der Sicht des Sammlers"* by C. Djerassi in PAUL KLEE FORMEN-SPIELE (ed. K. A. Schröder &Susanne Berchtold), Hatje Cantz, Ostfildern 2008, pp. 11-17.
*222. *"Vier Juden auf dem Parnass—ein Gespräch: Benjamin, Adorno, Scholem, Schönberg"* by C. Djerassi with illustrations by Gabriele Seethaler, (German translation by Ursula-Maria Mössner), Haymon Verlag, Innsbruck 2008, pp. 212.
*223. *"Il Seme di Menachem"* by C. Djerassi (Italian translation by Maria Pia Felici), Di Renzo Editore, Rome 2008, pp. 242.
224. *"Science on Stage: Why would a Chemist wish to write Plays?"* by C. Djerassi in *Canadian Chemical News,* Vol. 60, No. 4, April 2008, pp, 16-18.
225. *"Dopen, aber richtig"* by C. Djerassi in SÜDDEUTSCHE ZEITUNG, 21 May 2008, Nr. 125, p. 2.
*226. *"Sex in an Age of Technological Reproduction: ICSI and Taboos"* (with accompanying DVD) by C. Djerassi, University of Wisconsin Press, Madison 2008.
*227. *"Four Jews on Parnassus-a Conversation: Benjamin, Adorno, Scholem, Schönberg"* (with accompanying CD) by C. Djerassi (illustrations by Gabriele Seethaler), Columbia University Press, New York 2008, pp. 203.
228. *"Warum wir bald sehr alt ausschauen"* by C. Djerassi in DER STANDARD (Austria), 13 December 2008, Album A 3.
*229. *"Cómo derroté a.... y otros Cuentos"* by C. Djerassi (Spanish translation by Abdiel Macias), paperback edition, Fondo de Cultura Económica, Mexico, 2008, pp. 223.
230. " *Visual Culture and Bioscience—an Online Symposium"* (ed. Suzanne Anker and J.D. Talasek), *Issues in Cultural Theory 12,* publ. by University of Maryland and Cultural Programs of National Academy of Sciences, Washington DC, 2008, Contributions by C. Djerassi, pp. 24-29, 39-41, 69-70, 95-97, 146-147, 186-187, 193-195, 205.

2009

231. *"Was wäre, wenn es es die Pille nicht gäbe?* By C. Djerassi in *Süddeutsche Zeitung WISSEN,* June 2009, p. 91.
*232. *"Nach 70 Jahren: Wiener Amerikaner oder amerikanischer Wiener?"* By C. Djerassi, V&R unipress, Göttingen 2009, pp. 39.
233. *"From the Pill, Pygmy Chimps, and Degas' Horse"* by C. Djerassi in *Becoming Americans: Four centuries of Immigrant Writing* (ed. Ilan Stavans), The Library of America, New York 2009, distributed by Penguin-USA, pp. 304-312.

234. *"Ludwig Haberlandt – Grandfather of the Pill"* By C. Djerassi in *Wiener klinische Wochenschrift*, 121, 727–728 (2009).
235. *"Three on a Couch"* by C. Djerassi in "The Best Men's Stage Monologues and Scenes" (ed. Lawrence Harbison), Smith & Kraus Inc., Hanover, NH 2009, p. 84.

2010

*236. *"Cuatro Judíos en el Parnasso-Una Conversación: Benjamin, Adorno, Scholem, Schönberg"* by C. Djerassi with illustrations by Gabriele Seethaler, (Spanish translation by Cecilia Absatz), Capital Intelectual, Buenos Aires, 2010, pp. 213.
237. *"Die Antibaby Pille: Schuldig oder...?"* by C. Djerassi in "Katalog zur Ausstellung: die 60er: Beatles, Pille und Revolte," Schallaburg, Austria, Schallaburg Kulturbetriebsges. 2010, pp. 148-154.

2011

*238. *"Foreplay"* by C. Djerassi, University of Wisconsin Press, Madison 2011 (ISBN 978-0-299-2834-6).
*239. "Vorspiel" by C. Djerassi (German translation by Ursula-Maria Mössner), Haymon Verlag, Innsbruck 2011.
*240. "Preludio: una historia de sexo en la Escuela de Frankfurt" by C. Djerassi (Spanish translation by Cecilia Absatz), Capital Intelectual, Buenos Aires, 2011.
*241. *"Marx è morto "* by C. Djerassi (Italian translation by Francesca Garofoli), Di Renzo Editore, Rome 2011.
242. *"La Història d'l'Obra Teatral 'Oxigen': Ciencia i Literatura, del Paper a l'Escenari"* by C. Djerassi (Catalan translation)in MÈTODE (Univ. of Valencia, Catalan edition),Nr. 69, Spring issue, 105-110 (2011).
243. *"La Historia de la Obra Teatral 'Oxígeno': Ciència y Literatura, del Papel al Escenaroi"* by C. Djerassi (Spanish translation) in MÈTODE (Univ. of Valencia, Spanish edition), Nr... 69, Spring issue, 97-102 (2011).
244. *"The Pilll: no heir apparent"* by C. Djerassi in PROSPECT (UK) Sept. 2011, pp. 76-78.
245. *"Synthesizing a Life: An Interview with Carl Djerassi"* by Liberato Cardellini, J. Chem.Educ., 88, 1366–1375 (2011)
*246. *"Oxigen"* by C. Djerassi and R. Hoffmann (Catalan translation by Arantxa Gorostiza and Mercè Piqueras), MÈTODE, Valencia, 2011.
*247. *"Cálculo"* by C. Djerassi (Portuguese translation by Mário Montenegro), University of Coimbra Press, Coimbra 2011.
*248. *"Falácia"* (*Phallacy*) by C. Djerassi (Portuguese translation by Manuel João Monte) University of Porto Press, Porto, 2011.
249. *"99 Words"* by C. Djerassi in *"You have breath for no more than 99 words. What would they be?"* (Liz Gray, ed.), Darton, Longman & Todd, London, 2011 (ISBN 978-0-232-52889-3

2012

250. *"What can the theatre do for science and vice versa?"* by C. Djerassi in (Portuguese translation by Mário Montenegro) *"Sinais de cena"* 2012.
*251. *"Chemistry-in-Theatre: Insufficiency, Phallacy or Both?"* by C. Djerassi, Imperial College Press/World Scientific Publ., London and Singapore, 2012.
*252. *"Tagebuch des Grolls 1983-1984"* (*"A Diary of Pique 1983-1984,"*). A bilingual Poetry Collection (translated by Sabine Hübner) by C. Djerassi, Haymon Verlag, Innsbruck 2012

Contributors

CARL AIGNER studied History, German literature and Art History in Salzburg and Paris. He has been teaching at various universities since 1998, including the Universität für Angewandte Kunst in Vienna and the Donau-Universität Krems. Between 1997 and 2003 he served as director of the Kunsthalle Krems. In 2001 he became director of the Landesmuseum Niederösterreich. In 2002, he organized a special summer exhibition of Carl Djerassi's Klee collection in Krems – an event that eventually led to the Austrian government's decision to offer Djerassi the Austrian citizenship.

CARL DJERASSI published over 1200 scientific publications, which garnered him numerous honors, such as membership in the National Academy of Sciences, the Royal Society (London) and the Leopoldina (Germany), as well as 27 honorary doctorates. He is the only living chemist to have received both the National Medal of Science (for the first synthesis of a steroid oral contraceptive) and the National Medal of Technology. During his literary life, aside from two autobiographies, he has authored 5 novels, 9 plays, and 2 volumes of poetry, which collectively he calls a still ongoing form of auto-psychoanalysis.

ROBERT MARC FRIEDMAN is a professor of history of science and dramatist. His many publications as well as his four stage plays and two TV screenplays analyze modern science as a social, cultural, and cognitive enterprise. His scholarship and dramatizations have received international honors. In 2009 he was named Tetelman Fellow at Jonathan Edwards College, Yale University. He was befriended by Carl as a comrade-in-arms and is indebted to him for inspiration and encouragement.

INGRID GEHRKE currently works as head of International Relations at FH Joanneum University of Applied Sciences in Graz, Austria. In 2011 she received a Fulbright Schuman Scholarship for International Educators to the University of Minnesota. Ingrid Gehrke holds a degree in English, American Studies and German from Graz University and a Masters in Comparative

Literature from SUNY Binghamton. Her dissertation focused on the literary work of Carl Djerassi and was published in 2008 under the title *Der Intellektuelle Polygamist*.

ISABELLA GREGOR is a Viennese director who has directed plays, modern operas and operettas in major theatres and festivals in Germany, Austria, Switzerland, UK, USA and Singapore. Her collaboration with Carl Djerassi goes back to the first German theatrical premiere of his first play *An Immaculate Misconception*. This was followed by the first German premieres of *Oxygen* by Carl Djerassi and Roald Hoffmann, *Ego* by Carl Djerassi – in a special version written by Susanna Goldberg and herself – and *verrechnet!* written by Carl Djerassi and herself.

WALTER GRÜNZWEIG is professor of American Literature and Culture at TU Dortmund University and adjunct professor at the University of Pennsylvania, the State University of New York at Binghamton and Canisius College. In 2010, he was awarded the Ars Legendi Prize for excellence in university teaching by the German Rectors Conference. A native Austrian, he is interested in transatlantic cultural relationships. Carl Djerassi's life and work therefore interest him both in his research and on a personal level.

HANS ULRICH GUMBRECHT is the Albert Guerard Professor in Literature at Stanford University. His widely translated books include *Eine Geschichte der spanischen Literatur* (1990); *The Powers of Philology* (2003) and *In Praise of Athletic Beauty* (2006). Gumbrecht is a regular contributor to the *Frankfurter Allgemeine Zeitung* and the *Neue Zürcher Zeitung* and a member of the American Academy of Arts & Sciences. Over the past twenty-four years, has known his Stanford colleague Djerassi as an exceedingly generous friend and as a partner in a steady intellectual exchange.

HENNING HOPF is retired professor of organic chemistry who studied chemistry at Göttingen and at the University of Wisconsin-Madison where he received his Ph.D. in 1967. He taught at different German universities and concluded his teaching career at Braunschweig. He has known Carl Djerassi for decades both as chemist and writer and has reviewed his novels for German chemistry journals. He is most impressed by Djerassi's wit, his sometimes sarcastic humor and his insight into the *condition humaine*.

MARTIN JAY is Sidney Hellman Ehrman Professor of History at the University of California, Berkeley. Among his works are *The Dialectical Imagination* (1973 and 1996); *Marxism and Totality* (1984); *Adorno* (1984); *Permanent Exiles* (1985); *Cultural Semantics* (1998); *Refractions of Violence* (2003);

Songs of Experience (2004); *The Virtues of Mendacity: On Lying In Politics* (2010) and *Essays from the Edge* (2011). His interest in Carl Djerassi's work stems from a shared fascination with German émigré intellectuals.

ANDY JORDAN is a Senior Lecturer at The Lincoln School of Performing Arts at The University of Lincoln (UK). He also is a producer/director in theatre, working in London's West End, at the Edinburgh Festival, and abroad, as well as in radio, where he was a Senior BBC Radio Drama Producer, having directed numerous internationally award-winning radio productions. Over a period of 14 years, he has directed and produced all of Carl Djerassi's theatre plays in the UK as well as his two radio plays.

GEORGE KLEIN was born in Budapest and received his MD from the Karolinska Institutet where he served as Professor of Tumor Biology from 1957 until 1993 and since then as Research Group Leader. An internationally renowned scholar, he is a member of the Royal Swedish Academy of Sciences and numerous other international societies and acdemies. He has known Carl Djerassi since 1961 when he spent a sabbatical at Stanford. Their work in literature as well as well as their similar backgrounds provided many topics for discussions.

MATTHIAS KLEINER studied Mechanical Engineering at Universität Dortmund and received his Habilitation in the area of Forming Technology. From 1994-1998 he was Professor for Design and Manufacturing at Brandenburgische Technische Universität Cottbus. In 1997 he was awarded the Gottfried Wilhelm Leibniz Award of the German Research Foundation; one year later he became Professor for Forming Technology in Dortmund. In January 2007, Matthias Kleiner was elected President of the German Research Foundation (DFG); in 2011 he became co-chair of the German Ethics Commission for Secure Energy Supply. In his position of DFG president, he has repeatedly turned to Carl Djerassi for his interdisciplinary expertise.

PIERRE LASZLO was a professor of chemistry from 1966 to 1999 and witnessed Carl Djerassi's manifold accomplishments. He also taught French literature for a term at Johns Hopkins University. Upon retirement, he became a science writer and an historian of science. His latest published book in English is *Citrus: A History*, University of Chicago Press, 2007.

DAVID LODGE is a novelist, critic, and Emeritus Professor at the University of Birmingham where he taught for many years. His novels include *Small World* (1984), *Nice Work* (1988), *Thinks...* (2001), and most recently *A Man of Parts* (2011). He has also written plays and screenplays. He has been a personal friend of Carl Djerassi since they met in 1992 and shares his interest in communicating ideas in a variety of forms.

ERIN MCGLOTHLIN is Associate Professor of German and Jewish Studies at Washington University in St. Louis. Her research interests are in the areas of German-Jewish literature and Holocaust literature. She is the author of *Second-Generation Holocaust Literature: Legacies of Survival and Perpetration* (2006) and her articles have appeared in *Narrative*, *The German Quarterly*, *Gegenwartsliteratur* and other journals and edited volumes. She came to know Carl Djerassi's work in a course taught in Dortmund in 1998.

URSULA-MARIA MÖSSNER was born in Lübeck and studied English and Italian at the School of Translation, Interpreting, Linguistics and Cultural Studies of the University of Mainz in Germersheim. Following a stay in England between 1967 and 1972, she moved to Ulm and has since worked as a freelance translator (Truman Capote, Dorothy Parker, David Benioff, Maile Meloy). Since 1990 she has been the "German Voice" of Carl Djerassi, translating his novels, short stories, plays, and autobiographical works.

CHRISTIAN NOE studied chemistry and pharmacy. He was professor at TU Wien, then in Frankfurt and now at Vienna University and served as dean at the latter two universities. He was Founding President of the *Ignaz-Lieben-Society for Studies in the History of Science*. His main research interests are in the field of drug research. His interest in Carl Djerassi focuses on the scientific and professional achievements of this outstanding personality but is also related to his desire to welcome those who were driven from Vienna during the Nazi time back to their native city.

FEDERICO MARIA RUBINO of Milano is a trained chemist and a practitioner of mass spectrometry in biomedical research in Italy. He first encountered Djerassi's scientific work and later *Pygmy Chimps*. When he read *Cantor's Dilemma* he immediately wished to translate the novel into Italian and had it published in 2004. Has also rendered *The Clock Runs Backwards* poems into his native language and hopes to see it in print soon.

KLAUS ALBRECHT SCHRÖDER is an art historian, an exhibition maker and the director of the Albertina art museum in Vienna. This profile would not automatically provide for a relationship with one of the most important chemists of our time. However, their shared love for Paul Klee and Schröder's appreciation for Djerassi's never-ending curiosity, thoughtfulness as well as immense and infinite wisdom regarding life and the philosophy of life make for a strong affinity between the two.

GABRIELE SEETHALER studied biochemistry in Vienna and received her Ph.D. at the Institute of Molecular Biology of the Austrian Academy of Sciences in Salzburg. After the birth of her daughter Katharina in 1994, she started to fuse art and science through her widely exhibited "Identity genotype-phenotype" project. In 2006 Gabriele Seethaler began an intense collaboration with Carl Djerassi on *Four Jews on Parnassus – A Conversation* with exhibitions and projections of her work in Salzburg, Berlin, Stockholm, Las Palmas, Dortmund and Vienna.

EVA-SABINE ZEHELEIN is a member of the North American Studies Program at the University of Bonn. Her research interests are popular culture, the novel in the 21st century, American art and the natural sciences-the arts-interface. In 2009, her book *Science: Dramatic. Science Plays in America and Great Britain, 1990-2007* was published in which Carl Djerassi's plays feature prominently.

Index of Names

Aaserud, Finn 67
Abrams, M.H. 173, 181
Adams, John 10
Adorno, Theodor W. 10, 59, 196, 201, 210, 211, 219, 222, 224-228
Akin, Kathryn 125
Albrecht, James 125
Aldston, Alfred N., Jr. 166
Amis, Kingsley 10
Amundsen, Roald 73
Angier, Carole 159, 160, 166
Archer, Karen 125
Archimedes of Syracuse 33
Arden, John 100
Arendt, Hannah 59
Aristotle 16, 64
Augustine of Hippo 221

Balaram, Padmanabhan 77, 83
Barthes, Roland 148, 185, 190
Bartók, Béla 35
Beckett, Samuel 98
Bending, Jo 125
Benjamin, Walter 10, 59, 187-191, 194-197, 201, 210, 211, 214, 217, 219, 220, 224, 225
Bennett, Dan 111, 117, 121
Benzer, Seymour 33, 35
Bergmann 30
Berkowitz, Gerald 116, 121, 125
Bernhard, Thomas 46
Bevan, David G. 153
Bevell, Brian 112
Birtwistle, Helen 126
Bjerknes, Vilhelm 65
Blakemore, Michael 119
Blood, Diane 106
Boethius 221
Bohr, Margrethe 80
Bohr, Niels 94
Bond, Edward 100
Borges, Jorge Luis 12
Borzekowski, Dina L. G. 166
Bradbury, Malcolm 151
Brant, Sebastian 196
Brazier, Chris 125

Brecht, Bertolt 64, 89, 100
Brenner, Sidney 27, 31
Brown, Peter 125
Bryant, Nicola 125
Buber, Martin 223, 224
Buñuel, Luis 30
Bush, Vannevar 145f.

Campos, Liliane 115, 121
Carl XVI Gustaf of Sweden 57
Carlehed, Ingemar 70
Carter, Ian 153
Chaffie, Martin 162
Chargall, Marc 164
Cham, Jorge 162, 166
Chaucer, Geoffrey 105
Christo [Christo Vladimiroff Javacheff] 20
Cibber, Colley 113
Cicero, Marcus Tullius 221
Clark, Hamish 125
Coelho, Jorge, 162, 166
Cohen, I. Bernard 83
Cohen, Josh 125
Coleman, Toni 181
Colin, Martin 83
Copernicus, Nicolaus 221
Corner, Chris 112
Craig, Russell 111, 112, 123
Cram, Donald 149
Cricks, Francos 33, 41
Cronin, Archibald Joseph 165
Csíkszentmihály, Mihály 37-39, 41
Cusack, Catherine 123

Dante Alighieri, 105, 201
Darrow, Clarence 221
Davenport, Lucy 123
De Cruif, Paul 37
Demerger, Robert 123
Deppe, Renald 186
Derrida, Jacques 40, 148
Devins, Dorian 83, 99, 121
Diner, Dan 218, 219
Djerassi, Alice [Alice Friedmann] 21
Djerassi, Pamela 20, 171

Djerassi, Samuel 20f.
Donne, John 173m 174, 175
Donovan, Arthur 83
Dürrenmatt, Friedrich 65, 195
Dunnett, Iain 122
Duse, Eleanora 89

Eberlein, Konrad 187
Edwards, Lynette 124, 125
Ehrenreich, Barbara 169
Einstein, Albert 76, 105
Eisen, Jonathan A. 166
Eliot, Thomas Stearns 157, 163
Erasmus of Rotterdam 221

Fansler, Kate 151
Feigenbaum, Edward 57
Fenner, Michael 124
Fink, Sheri L. 166
Fitzgerald, Geraldine 123
Fleming, Ian 159, 167
Foucault, Michel 40, 148
Frayn, Michael 80, 83, 93, 94, 102, 115, 119
Freud, Sigmund 21, 105, 221
Frisch, Otto 70
Fujimura, Joan 40

Galileo Galilei 14, 221
Gant, David 124
Gardner, Helen 173, 174, 179, 181, 182
Gay, Peter 222
Gehrke (née Gomboz), Ingrid 5, 143
Genzlinger, Neil 4
Godwin, Gail 151
Goethe, Johann Wolfgang von 43, 44, 46, 159, 213, 214, 217, 220
Goldstein. Moritz 216-218, 220
Goodwin, Paul 123
Gräslund, Astrid 82
Green, Chris 106
Greenlaw, Lavinia 112
Greer, Germaine 122
Gregor, Isabella 96, 112, 123
Grotowski, Jerzy 89
Gryphius, Andreas 215
Gunzenhäuser, Randi 4

Gustav III of Sweden 76, 81

Haber, Fritz 76
Haberlandt, Ludwig 54
Franz Stephan Habsburg-Lothringen 53
Händel, Georg Friedrich 201
Haffmans, Gerd 10
Hahn, Otto 69
Halliburton, Rachel 111, 121, 123
Hare, David 100
Hargittai, István 36
Hartmann, Geoffrey H. 187
Hegel, Georg Wilhelm Friedrich 138, 228
Heilbrun, Carolyn 151
Heine, Heinrich 214, 216
Heisenberg, Werner 80, 94
Hensarling, Jeb 160
Hemming, Sarah 76, 83
Heraclitus of Ephesus 197
Herder, Johann Gottfried 215
Herzl, Theodor 217
Hesse, Hermann 164
Heymann, Inger 70
Hickman, Julia 116, 121, 123
Hitler, Adolf 187, 189
Hjalmarsson, Barbro 82
Hoen, E. Weber 166
Hoffmann, Roald, 28, 46, 58, 75-83, 92, 109, 110, 112, 113, 117, 159, 167
Homer 221
Hooper, Warren 125
Hoyle, Fred 151
Hubball, Debra 127, 129-131
Hughes, Howard 221
Hughes, Kevin 98, 121, 124
Hughes, Ted
Hung, Dean Y. 166
Hussein, Saddam 221

Iachello, Francesco 72

Jack the Ripper 221
Jacob, François 27, 31
Jacoby, Daniel 214
Jay, Martin 226, 229
Jeanne-Claude [Jeanne-Claude Denat de Guillebon] 20

Jelinek, Elfriede 86
Johns, Ian 124, 126
Jones, Beaudelaire 221, 229
Jonson, Björn 72
Jordan, Andy 122-125
Joyce, James 152

Kallstenius, Edvin 82
Kane, John 124, 125
Karlberg, Johan 70
Kasparov, Gary 15
Kauffmann, George B. 110, 112
Kauffmann, Laurie N. 110, 122
Kekulé, August 43
Kipphardt, Heinar 165
Klaff, Jack 123, 125
Klee, Paul 10, 18, 157, 187, 190, 191, 193-197, 201, 204
Klein, Eva 29
Klopstock, Friedrich Gottlieb 215
Kluge, P.F. 151
Kodály, Zoltán 35
König, Burghard 44
Kortner, Fritz 89
Kuhn, Thomas 77, 83

Lacan, Jacques 148
Landauer, Gustav 224
Lavoisier, Antoine Laurent 75-78, 81, 92, 93, 109, 110, 112, 148
Lavoisier, Marie-Anne 79, 80, 81, 83, 91, 112
Leconte de Lisle, Charles Marie René 222
Lederberg, Joshua 37, 57
Leibniz, Gottfried Wilhelm 59, 114
Lenin [Vladimir Ilyich Ulyanov] 105
Lessing, Gotthold Ephraim 215
Lethen, Helmut 226
Levi, Primo 158, 159, 167
Liemann, Lucy 125
Lin, Shirley 166
Lodge, David 151
Lubitsch, Ernst 89
Lucian of Samosata 221, 229
Lurie, Alison 151
Lynch, Suzanne 115, 117, 122

Machiavelli, Niccolò 19

Mann, Golo 195
Maria Theresia of Habsburg 53
Martyn, Rhonda 10
Marx, Karl 105
Masters, Edgar Lee 159
May, Roger 124
McCarthy, Mary 10, 151
McClintock, Barbara 33-35
McCormick, Katharine Dexter 54
Mead, Margaret 137
Mendel 82
Mengs, Anton Raphael 213
Menippus 221
Métraux, Rhonda 137
Meitner, Lise 68, 69, 70
Merton, Robert 39
Minckwitz, Johannes 215, 216, 220
Meurer, Petra 4
Middlebrook, Diane 9, 10, 11, 13, 20, 35, 38, 58, 85, 157, 169-172
Milgram, Jennifer 117, 122
Mill, John Stuart 12
Miller, Arthur 162
Moore, Michael 221
Moriarty, Paul 122
Moseley, Merritt 153
Mountford, Fiona 126
Mullis, Kari 36, 37
Murphy, Jack 112, 123
Musil, Robert 43

Nabokov, Vladikmir 151
Nansen, Fridtjof 73
Napoleon Bonaparte 105
Neill, Heather 116, 120, 122
Neuhuber, Franz 186
Newton, Isaac 59, 82, 105, 113, 114
Nguyen, Cynthia T.M.H. 166
Nicholls, Jon 125

Olivares, Marcela 125
Opitz, Martin 215
Oxman, Steven 79, 83

Pallavicini, Piersandro 162, 167
Parker, Joanna 107, 113, 122

Perry, Jane 125
Pessoa, Fernando 201
Phillips, Julie E. 166
Picasso, Pablo 53, 58, 164
Pinter, Harold 67, 97, 98
Pirandello, Luigi 64, 164
Plath, Sylvia 169
Plato 196, 221
Plessner, Helmut 226
Plutarch 221
Poivier, Jean-Pierre 83
Ponzio, Giacomo 160
Popescu, Lucy 124
Prasher, Douglas 163, 167
Prelog, Vlado 149
Price, Andy 112, 129
Prieler, Claudia 86
Priestley, Joseph 75, 77, 78, 81, 92, 93, 109, 110, 164
Priestley, Mary 79, 80, 81
Pym, Barbara 151

Raphael 213
Rayl, A.J.S. 110, 111
Reinhardt, Max 89
Richter, Gerhard 224f., 229
Rimbaud, Arthur 186, 191
Rist, Johann 211, 212, 217, 220
Rivers, Isabel 83
Rockefeller, John D. 221
Rohn, Jennifer 117, 122, 125, 165, 167
Rosenqvist, Helena Hillar 82
Ross, Alex 229
Rubino, Letizia Morabito 166
Rushton, Annie 128
Russo, Richard 151

Sagredo, Giovanni Francesco 221
Sandiland, Nic 107, 122
Sanger, Margret 54
Scheele, Carl Wilhelm, 75-79, 81, 82, 92, 93, 109, 110, 164
Scheele, Sara Margarethe [Fru Pohl] 79, 80, 83
Schiebinger, Londa 137-139
Schiller, Friedrich von 64, 215
Schlegel, August Wilhelm 215
Schlegel, Friedrich 215

Schneider, Adolf Wilhelm 215, 220
Schoen, Ernst 225
Schönberg, Anton 10, 194-197, 201, 210, 211, 219
Schönberg, Mathilde [Mathilde Zemlinsky] 222
Schofield, Robert E. 83
Scholem, Gershom 10, 196, 201, 210, 211, 222, 214, 217-220, 222, 223, 224, 229
Schulze, Werner 124
Seethaler, Gabriele 10, 185-192
Shakespeare, William 64, 82, 98, 105
Sheppard-Barr, Kristin 101, 102, 115
Sheridan, Susan 124
Shimomura, Osamu 162
Shore, Robert 124
Showalter, Elaine 153
Siegbahn, Manne 69, 70
Smiley, Jane 151
Smith, Anna Nicole 221
Snow, Charles Percy 6, 11, 26, 117, 142, 151
Socrates 196
Stephenson, Debra 122
Stephenson, Shelagh 75, 83
Sterne, Laurence 196
Stevens, Wallace 13, 157, 163, 178f.
Stohlmeyer, Michelle 166
Stoppard, Tom 93, 97, 98
Strasberg Lee 88, 89
Stravinsky, Igor 227
Sumen, Cenk 166
Svanholm, Set 82
Swansen, Craig A. 141-143

Tagiguchi, Norika 166
Taylor, Bayard 214, 220
Taylor, Jeremy Lindsay 125
Taylor, Michael 115, 124, 125
Thain, Paul 98
Thaxter, John 114, 122-124
Thiessen, Vern 75, 76, 83
Thompson, Helena 124
Thorstenson, Yvonne 166
Tiedge, Christoph August 215
Tomatis, Renzo 162, 166
Tsien, Roger 162
Turnstall, Darren 100, 124, 125

Tyler, Christian 122

Unwin, Paul 98

Vanbrugh, Sir John 113
Verdi, Guiseppe 81
Voltaire [François Marie Arouet] 221

Warhol, Andy 196
Washington, Harriett A. 166
Watson, James D. 26, 27, 31, 41
Weichmann, Ludwig 65
Whyman, Eric 112
Wieland, Christoph Martin 213, 215, 220
Wilde, Oscar 196
Wilder, Billy 89
Williams, Dudley 159, 167
Willingham, Tyrone 15
Wilton, Terence 122, 124
Winckelmann, Johann Joachim 212, 213
Winograd, Terry 15
Womack, Kenneth 153
Woodward, Robert 65
Wordsworth, William 12
Wykes, David L. 83

Vendl, Alfred 58

Zehelein, Eva-Sabine 5, 102, 115, 119, 122
Zimmermann, Mary 87
Zorn, Anders Leonard 82

Index of Works by Carl Djerassi

The Bourbaki Gambit 28, 30, 32, 33, 58, 136, 139-142, 152, 153, 163, 164, 201, 205
Calculus 39, 59, 104, *113-116,* 118, 124, 129, 204
Cantor's Dilemma 3, 4, 10-14, 16, 28, 33, 36, 42, 51, 58, 135, 136, 138-142, 151, 160, 162-166, 199-205
The Clock Runs Backward 157-159, 161, 169, 178-179, 181
A Diary of Pique 169 – 181
Ego (Three on a Couch) 40, 91, 104, 118, 201, 205
Foreplay 104, 121, 205
Four Jews on Parnassus 5, 9, 10, 14, 34, 59, 89, 90, 143, 193 – 197, 201, 205, 209-211, 218-222, 225f., 227
The Futurist 13, 177, 204, 205
An Immaculate Misconception 14, 58, 80, 91, 92, 98, 104, *105-107, 122-123, 127*, 201, 204
Marx, Deceased 104, 152, 163
Menachem's Seed 58, 201, 202f., 205
NO 40, 58, 140, 141, 201, 203
Oxygen 14, 28, 33, 39, 42, 75 – 83, 91, 104, *109-113,* 116, 118, 123 – 124, 128, 164, 204
Phallacy 58, 87, 91, 116 – 117, 118, 125, 130, 204, 205
The Pill, Pygmy Chimps, and Degas' Horse 10, 28, 31, 40, 58, 200, 204
Steroids Made It Possible 27, 40, 161
Taboos 4, 104, 107-109, 125 – 126, 131, 201, 205
This Man's Pill 205

Transnational and Transatlantic American Studies
edited by Kornelia Freitag (Bochum), Walter Grünzweig (Dortmund), Randi Gunzenhäuser (Dortmund), Martina Pfeiler (Dortmund) Wilfried Raussert (Bielefeld), Michael Wala (Bochum)

Ingrid Gehrke
Der intellektuelle Polygamist
Carl Djerassis Grenzgänge in Autobiographie, Roman und Drama
Bekannt wurde er als der Wissenschaftler, dem wir die Pille verdanken.
Mit über 60 Jahren wechselte Carl Djerassi zur Literatur und schrieb Romane und Dramen über Naturwissenschaft und Naturwissenschaftler, die er als eigene Stammeskultur sah.
Sein faszinierendes Leben reflektiert er in einer Reihe von Autobiographien, die zeigen, wie sehr er das vergangene Jahrhundert mitgeprägt hat.
Das Buch liefert die erste Darstellung seines literarischen Werkes und verfolgt Djerassis Entwicklung zum Schriftsteller sowie seine Spurensuche nach seinen europäischen Wurzeln.
Bd. 9, 2008, 240 S., 29,90 €, br., ISBN 978-3-8258-1444-1

LIT Verlag Berlin – Münster – Wien – Zürich – London
Auslieferung Deutschland / Österreich / Schweiz: siehe Impressumsseite

Heike Steinhoff
Queer Buccaneers
(De)Constructing Boundaries in the PIRATES OF THE CARIBBEAN Film Series
Pirates captivate the western cultural imagination at the beginning of the 21st century. Queer Buccaneers addresses this phenomenon through an analysis of the Disney film series Pirates of the Caribbean. Reading the films from a variety of post-structuralist perspectives, this study demonstrates the contradictory discourses and power relations that characterize the series. It argues that 'piracy' constitutes a sliding signifier that facilitates the (de)construction of discursive boundaries of gender, sexuality, race, ethnicity, class and nationality.
Bd. 10, 2011, 152 S., 19,90 €, br., ISBN 978-3-643-11100-5

LIT Verlag Berlin – Münster – Wien – Zürich – London
Auslieferung Deutschland / Österreich / Schweiz: siehe Impressumsseite